Worl[d]

NUMB[ER]

## DO NOT REMOVE
## CARDS FROM POCKET

# The Economics of Afforestation

## A Case Study in Africa

*Dennis Anderson*

Published for The World Bank

THE JOHNS HOPKINS UNIVERSITY PRESS

Baltimore and London

The Johns Hopkins University Press
Baltimore, Maryland 21211, U.S.A.

The findings, interpretations, and conclusions expressed in this study are the results of research supported by the World Bank, but they are entirely those of the
author and should not be attributed in any manner to the World Bank, to its affiliat-
ed organizations, or to members of its Board of Executive Directors or the
countries they represent. The map that accompanies the text has been prepared
solely for the convenience of readers; the designations and presentation of material in it does not imply the expression of any opinion whatsoever on the part of the
World Bank, its affiliates, or its Board or member countries concerning the legal
status of any country, territory, city, or area, or of the authorities thereof, or concerning the delimitation of its boundaries or its national affiliation.

*Library of Congress Cataloging-in-Publication Data*

Anderson, Dennis, 1937–
    The economics of afforestation.

    (World Bank occasional papers; new ser., no. 1)
    Bibliography: p.
    1. Afforestation—Economic aspects—Africa,
Sub-Saharan. 2. Afforestation—Economic aspects—
Nigeria, Northern. 3. Deforestation—Africa, Sub-Saharan.
4. Deforestation—Nigeria, Northern. 5. Afforestation—
Government policy—Africa, Sub-Saharan. 6. Windbreaks,
shelterbelts, etc.—Nigeria, Northern. 7. Wood-lots—
Nigeria, Northern. I. Title. II. Series.
SD409.A58  1987        338.1'749'0967        87–45410
ISBN 0-8018-3552-6

# Contents

# Preface

EROSION AND LOSS OF SOIL FERTILITY arising from the destruction of trees over large areas of developing countries are among the most important problems of economic development today. The technical solutions have been well known for many years; they include the planting of public shelterbelts and the adoption of farm forestry practices—the maintenance of windbreaks, woodlots, small stands of trees, and farm trees and shrubs as part of farm management. But what is less appreciated, except by foresters and agriculturalists, is that investments in shelterbelts and farm forestry can be highly productive. They not only prevent deterioration of the soil but may actually enhance soil fertility, raise the output of cropping and livestock activities, and help to alleviate the scarcity of fuelwood.

Part I reviews the economic and institutional aspects of shelterbelts and farm forestry with special reference to Africa, discusses why farm forestry practices have so far not been widely adopted in the region, and makes policy proposals. Part II, which contains a case study of the arid zone of northern Nigeria, shows how the ecological benefits of rural afforestation programs can be translated into economic terms and demonstrates that such programs would be not only ecologically beneficial to the region but very likely economically beneficial as well. This conclusion is relevant for other arable areas that currently face ecological threats.

Part I is based on an earlier study with Robert Fishwick and on my 1986 article in *World Development*. It is reproduced here, with minor editing and additional reference material, by courtesy of Pergamon Press. Part II is the result of an extensive field trip to Nigeria in May and June of 1985 to assess previous afforestation programs in the region and the prospects for future programs. I am grateful to Ali Abibu, Olare Amosum, Rex Aroufar, Janus Debski, Peter Mamsa, Luke Umeh, and the chief conservators of forests and natural resources of Sokoto, Kano, Kaduna, Bauchi, and Borno states for many discussions and much insight provided during the field trip.

I have also benefited from many discussions on the subject with Abderraouf Bouhaouala, Stephen Carr, Stephen Eccles, Walter Elkan, Robert Fishwick, Hans Fuchs (who initiated the work that led to the book), Jean Gorse, Robert Hindle, Tariq Husain, Henrik Hvidberg-Hansen, Kenneth Newcombe, Donald Pickering, Gunter Schramm, Kenneth Sigrist, John Spears, Hugh Speechly, Roger Sullivan, and Jeremy Warford. David Pearce reviewed the economic analysis, and both he and Kenneth Sigrist offered comments that helped to tighten up the analysis. I appreciated their encouragement, as I did that of Norman Myers, who is always stimulating to talk to on ecological issues. Frank Thompson kindly arranged for me to give a talk on afforestation to postgraduate students at the Commonwealth Forestry Institute in Oxford; again, constructive comments and ideas emerged. Four anonymous reviewers of the manuscript also went beyond the call of duty in providing detailed and helpful comments and in drawing additional research material to my notice. Ghazali Raheem did a splendid job in writing the computer programs for the analysis, as did Sonja Motz in typing the manuscript.

The views expressed in the book and, of course, any defects in it are solely my responsibility. But the immense experience of the people I have just thanked gives me some confidence in thinking that the analysis and conclusions are on the right track.

# Abbreviations

AZAP    Arid Zone Afforestation Program
ILCA    International Livestock Centre for Africa
IRR    Internal rate of return
l.c.f.    Local conversion factor
MAI    Mean annual increment
NPV    Net present value
PV    Present value
s.c.f.    Standard conversion factor
TLU    Tropical livestock unit

# Introduction

SCIENTISTS, AGRONOMISTS, AND OTHERS have drawn attention to the serious and mounting ecological problems caused by deforestation and the loss of trees on farmlands and in watersheds in Africa. These concerns date back many years—even before Stebbing's paper in 1935 on the "encroaching Sahara." Despite shortcomings in the available evidence, it is widely agreed that large and increasing losses occur each year because of agricultural land clearing, commercial logging, and the harvesting of trees for fuelwood, fodder, and building materials. In many respects the loss of trees is not surprising, given population growth and the increasing demand for arable land, fuelwood, and timber. Indeed, there are appreciable short-run benefits from the value of the harvested wood and from economic output on lands cleared for agriculture. The cause of concern is that such benefits are being accompanied by huge and readily visible ecological costs, as measured by the long-run threat to the carrying capacity or fertility of the soil over large areas.

Why are deforestation and losses of trees on farmland occurring, and what can be done about it? More specifically, how can public policies best be brought to bear on the problem? The first part of this book (chapters 1 and 2) addresses these questions in a general, nontechnical way; the second part (chapters 3–5) presents a case study of the arid zone in northern Nigeria. The book concentrates on the loss of trees in farming areas, including woodlands and forests that are being cleared for agriculture, and it deals in depth with the economics of rural afforestation. It does not consider deforestation from commercial logging, which raises important but different and complex issues that require separate treatment.[1]

By rural afforestation is meant the maintenance or restoration of trees in farming areas—on farm boundaries, in villages and hamlets, near dwellings, in copses, in village woodlots, and in watersheds and shelterbelts. All such plantings, if carried out over large areas by farmers and forestry services together, protect soils from erosion and from loss of nutrients and moisture. Indeed, in

*1*

the more denuded areas plantings may enhance soil fertility. In areas still being cleared for agriculture the same ecological effects can be achieved at a fraction of the cost by leaving trees standing in the arrangements described above. The overall outcome would be an increase in farm incomes because of the higher outputs of crops and livestock; the effect would also be sustainable because the long-run threat to the soil's carrying capacity from erosion and from loss of nutrients and moisture would be reduced. In addition, there would be economically important by-products such as firewood, poles, fruit, mulch, and fodder.

Despite such often-discussed benefits of rural afforestation,[2] few attempts have been made to assess their economic value. With exceptions, the economic analysis of foresters tends to concentrate on wood, that of agriculturalists on crops, and that of livestock specialists on meat and dairy products, with no attempt to draw out the mutually beneficial associations between silviculture, agriculture, and livestock in rural areas. These associations have been more thoroughly discussed in ecological terms than in economic terms.[3] Such compartmentalization of analysis greatly understates what might be achieved, leads to misleading perspectives on public policy, and helps to explain why rural afforestation has been given little or no attention in rural development programs in the region. As a result of this inattention, public and private investment in agriculture is proceeding alongside private disinvestment in soil and may ultimately be undermined by the latter.

Another reason why a more complete analysis of costs and benefits has not been undertaken may be that quantitative evidence on the ecological changes that are taking place is seriously deficient, which in turn makes the quantitative analysis of the benefits of afforestation uncertain. Nevertheless, and however approximate the results, economic analysis on the basis of current knowledge and thinking can be useful for defining—and making the case for—public policy toward afforestation, including direct investment in the planting and protection of trees, in the provision of supporting services, and in research.

This book has four main purposes:

- To provide a brief overview of the extent and causes of the loss of trees in farming areas in Sub-Saharan Africa (chapter 1)
- To review policy options and outline an approach for evaluating afforestation investments (chapter 2)

- To demonstrate, using a case study in northern Nigeria, the usefulness of the approach for the economic analysis of afforestation projects in general (chapter 3–5)
- To propose topics for research on the effects of rural afforestation (chapter 5).

The main findings of the book can be summarized as follows. The technical solutions to the ecological problems that arise from the loss of trees on farmlands are well known; they include (not surprisingly) the replenishment and maintenance of farm tree stocks and the adoption of soil and farm management practices that would sustain—and probably enhance—the carrying capacity or fertility of the soil. Burley (1982, p. 41) remarks, "within arid and semi-arid lands . . . there are large differences in climatic and edaphic factors, but . . . these are not the factors limiting tree planting. Drought-hardy and salt-tolerant species exist that can be planted profitably and yield a range of end products."

As subsequent chapters show, knowledge about the design and implementation of afforestation programs has grown. Much that is constructive can be learned from successes in India, China, and other places and even from failures, and these lessons corroborate current thinking about the ingredients of a successful program.

One of the main conclusions is that afforestation programs have low costs per hectare of farmland protected. When their effects on soil fertility are allowed for, these programs could do much to improve the value of farm output—that is, they have prospects of good rates of return. As with all investments in low-income and ecologically threatened regions, there are significant risks, mainly because of uncertainty about how farmers will respond to the programs. But it seems to me that the ecological risks of doing nothing far outweigh the financial risks of proceeding with rural afforestation programs. Indeed, this is one way of interpreting the results of the cost-benefit analysis described in part II. (The research suggested in chapter 5 is intended in part to help reduce the risks.) A positive commitment to rural afforestation and to improving soil and farm management practices, combined with better incentives for investment in agriculture, could successfully address the serious ecological problems that now threaten large tracts of arable land.

## Notes

1. See, for example, World Resources Institute, World Bank, and United Nations Development Programme (1985), Myers (1984), Lanly (1982), and World Bank (1978).

2. See, for example, Food and Agriculture Organization (1985), Foley and Barnard (1984), National Research Council (1984), Wiersum (1984a), Mnzava (1983), Burley (1982), and von Maydell (1979).

3. For example, in the references given in note 2.

# PART I

## Deforestation in Africa: Evidence and Policies

CHAPTER 1

# Fuelwood Consumption and Land Clearing

IN AFRICA around 90 percent of the population use fuelwood for cooking. Measured in oil equivalent units, the amount is roughly 1.5 tons of oil a family a year.[1] In low-income countries the consumption of fuelwood energy by households is typically ten times the total consumption of commercial energy for all purposes, including transport and the generation of electricity; in Nigeria it is twice the total. The current annual rate of consumption is esti-. mated to exceed the mean annual incremental growth (generally called mean annual increment, or MAI) of local tree stocks and forest reserves by the following (rounded) amounts: in Senegal −35 percent (a slight surplus), in the Sahelian countries 30 percent, in Sudan 70 percent, in northern Nigeria 75 percent, in Ethiopia 150 percent, and in Niger 200 percent.[2] Although the estimates are not precise (much fieldwork is needed to obtain better measures of the MAI in practically all well-populated regions of Africa), they do reflect the situation in many regions. They neglect, however, the effects of land clearing, which many believe contributes more to the loss of trees than does fuelwood consumption.

Once consumption exceeds the MAI in a region, any or all of three things can happen to tree stocks: the deficit can be partly made up by woodlots and forestry plantations, which helps to keep overall stocks intact; natural stocks may be harvested locally, which increases the gap between consumption and the MAI and reduces stocks further; or natural stocks may be harvested elsewhere, which provides some respite for the fuelwood-deficit areas but spreads deforestation to other regions. In practice, private and public plantations make insignificantly small contributions to the MAI—less than 5 percent in most areas—and it is the felling

*Note:* Chapters 1 and 2 are reprinted, with minor editorial changes, from Dennis Anderson, "Declining Tree Stocks in African Countries," *World Development*, vol. 14, no. 7 (1986), pp. 853–63. © 1986 Pergamon Journals, Ltd.

of naturally growing trees and the spread of deforestation that are most noticeable.

Furthermore, the growth of deficits within an area, or their spread to other areas, accelerates because consumption increases exponentially while the MAI declines in proportion to the volume of stocks. There are simple dynamic equations which capture these effects clearly.[3] They show that for defined areas with deficits of, say, about 20 percent the rate of loss of local tree cover may typically be only around 2 percent a year—a theoretical lifetime for local tree stocks of forty years or more if felling is confined. Under the influences described, however, the rate of decline triples in two decades, and the lifetime of the local stocks is closer to twenty years. Such estimates allow for price effects and for the substitution of commercial fuels for fuelwood and charcoal but not for the effects of land clearing or for the decline of the MAI in many regions owing to the loss of seedlings and young trees to cattle.

The spread of deforestation is most noticeable near urban areas. The growth of towns and cities brings about appreciable demands for fuelwood and charcoal and accounts for much of the observed decline in tree stocks in the surrounding countryside, often for a radius of 80–160 kilometers or more. As long as local wooded areas and forests are not depleted, the cost of cooking with fuelwood or charcoal continues to be lower than the cost of the commercial alternatives. Fuller (1981) found that the total costs of collecting fuelwood from indigenous woodlands, transporting it to urban areas, and retailing were competitive with the costs of commercial fuels up to a radius of about 100 kilometers; for fuelwood plantations the costs (including plantation costs) were competitive up to a radius of 65 to 85 kilometers.[4] Such comparisons naturally depend on location and the quality of local roads and transport; conversion to charcoal further reduces transport costs and perhaps widens the radius. But the main point is that as long as local fuelwood can be supplied by traders urban areas burn it and in so doing greatly deplete the surrounding tree cover. In Ethiopia Newcombe (1984) found that once the local fuelwood resources were depleted an active urban market for animal dung developed. Thus the demand for fuelwood in urban areas remains strong and accounts for much of the spread of deforestation in Africa, notwithstanding the opportunities for substitution of commercial fuels in these areas.

With regard to land clearing by farmers, there is little that can be said in quantitative terms. As Timberlake (1985) notes, little good information has been compiled on what is taking place, aside from numerous but mostly unpublished field reports by foresters and others.[5] Further, most of the published data refer to forests and do not include the cutting of trees on farmlands and woodlands. As with many other aspects of scientific research in Africa, the analyst is soon confronted with the stark fact that information is deplorably lacking in quality and quantity. Nevertheless, it is not disputed that land clearing is taking place extensively and perhaps contributes as much as or more than fuelwood consumption to the depletion of stocks (which makes what was said earlier err on the side of underestimation). For example, Timberlake (1985, p. 106) reports that "population growth plays its part, as does migration into the coastal countries. An estimated five million hectares of forest were converted to agriculture in the Ivory Coast between 1966 and 1980; farmers destroyed some 300 million cubic meters of saleable timber, far more than was exported during the same period."[6] Felling of trees on farmlands is also widespread. Unpublished surveys in the arid northern zones of Nigeria, for instance, report that farm tree densities have declined from 15 to 3 a hectare since the 1950s. My observations from a field trip support these data: some newly cleared farmlands in Borno State are completely bare, with clearing being further encouraged by mechanization. Despite governmental attempts to encourage the planting of farm trees, the practice has been limited, with the notable exceptions of fruit trees, such as cashews and mangos, and trees for shade, such as neem. The overall picture is that the loss of trees is as widespread on farms as in forests and woodlands.[7]

That tree stocks can be expected to decline is not surprising, given the population-induced growth of demand for fuelwood and for land, and despite the poor quality of the data, the correlations are readily established. As Allen and Barnes (1985) point out, growth of per capita income might be expected to encourage a shift to commercial fuels, since fuelwood is an inferior good (in the economist's sense of the term), but in low-income countries the effect is apparently weak even where per capita incomes are increasing. Rising scarcity prices must eventually encourage substitution, but they also encourage a supply response—the harvesting of trees in surplus areas for sale in deficit areas. (For low-income farm families, this is a remunerative slack-season activity.) Whatever responses fuelwood consumption might show to changes in incomes and prices, however, the problem of land clearing remains.

Bringing more land under cultivation is as necessary as food production itself and, given the reality of population growth, it is difficult to see how it can be avoided. But the question is, why is land clearing taking place in such an ecologically costly way, as measured by a decline in soil fertility or even by losses of soil? The degradation is occurring not only on newly cleared land but also, as one might infer from the declines in farm trees, on land that has long been under cultivation. The well-known farm forestry practice of maintaining trees and shrubs on farm boundaries, near dwellings, and in copses could do much to protect soil fertility and could also supply fuelwood, fruit, and fodder. (See, for example, National Research Council 1984a, b; Myers 1984; and Foley and Barnard 1984.) But there are numerous reports that the rapid depletion of trees is causing unprecedented—and, without remedial policies, irreversible—changes in the ecology of rural areas, principally in the carrying capacity or fertility of the soil.[8] Although the nature, rate, and extent of the changes taking place vary appreciably between regions, four effects, and sometimes five, are commonly associated with the loss of trees: (a) gully erosion and the loss of topsoil to wind and rain; (b) greater surface evaporation and reduced moisture content of the soil as surface wind velocities increase; (c) greater surface runoff and adverse changes in water tables (which also increase the stress on existing trees); (d) a general reduction in the recycling of nutrients; and in some regions (e) the consumption of soil nutrients (dung and crop residues) as fuel when fuelwood becomes scarce. Newcombe's (1984) study in Ethiopia notes that the last effect is usually the terminal stage of the desertification process and can set in very quickly once fuel becomes scarce.[9] All these effects are in addition to the much-discussed problems of overgrazing, poor cropping practices, and the loss of young trees and seedlings to ill-managed livestock populations (leading to a lower MAI).

To return, then, to the above question: why are trees in farming areas being depleted in a manner that causes such common and visible ecological damage? At first glance it is paradoxical, since the farming communities themselves must ultimately bear most of the cost. Several explanations have been put forward, of which four will be noted here. They are not mutually exclusive, and in practice the problem may be to give the proper weight to each (and to other factors I may have omitted).

1. The most frequently cited explanation supposes a "commons problem" in Africa: communal tenure systems are said to under-mine the individual's incentive to invest in or even maintain land. The argument extends to the harvesting of trees on common woodlands and on unprotected state forest reserves to supply fuelwood-deficit areas. Important as this problem may be, how-ever, Noronha (1985, ch. 7), Famoriyo (1979), and others have cautioned that the term "communal tenure" is frequently ill-defined and that its close identification with "traditional tenure" in Africa is misleading. Tenure systems are complex, vary greatly, and are changing in many places under the influence of popula-tion growth, migration, economic change, and changes in land values. There are well-documented instances in which land under communal tenure has been sold or rented, and as Fortmann (1985) and others have observed, tree planting by private individu-als may permit them to establish private, although nonpermanent, claims to communal land. Noronha cites earlier studies of land tenure which conclude that people farming land in accordance with custom rarely consider themselves as lacking security; culti-vators generally possess land as long as it is cultivated, and their heirs frequently inherit rights to cultivate it. In some instances new laws and governmental intervention have undermined rather than enhanced the cultivators' sense of security and their incen-tive to invest in and maintain the land (Noronha 1985; Famoriyo 1979). Scholars are increasingly questioning the "tragedy of the commons" hypothesis, on theoretical grounds (the hypothesis assumes no cooperative behavior and no emergence of local lead-ership to manage communal affairs) and because the empirical evidence is weak (see Griffin 1986a; Livingstone 1986; Runge 1986). It would seem, then, that any defects in the changing sys-tems of tenure in Africa are only part of the explanation.[10]

2. It is often suggested that there are time lags in people's per-ceptions about soil erosion and loss of soil fertility. Related to this is the suggestion that the ecological value of trees may not be appreciated by farmers; in areas where trees were previously an abundant natural resource, farmers take them for granted. It is difficult to find good empirical evidence on these points, but again, in view of the ecological damage now taking place, it would be unwise to dismiss these possibilities.

3. Another explanation cites the "bias" against agriculture in the

macroeconomic and investment incentive policies of most African governments and the low priority given to agricultural development and afforestation in public expenditure programs. These issues have been thoroughly discussed elsewhere and require little elaboration here. (World Bank 1981 contains an extensive discussion and references.) Besides depressing the incentive to invest in agriculture itself, such policies have other consequences. They increase the demand for fuelwood in rural areas (low incomes raise the demand for an inferior good) and in urban areas (low rural wages mean low supply costs), drive people to harvest and market fuelwood from common woodlands and even dung from arable lands to derive supplementary income from slack-season labor, and, it is thought, cause farmers to discount future ecological costs at a high rate.

4. Finally, there is the matter of external costs. The most commonly discussed case is that of watersheds: the costs of silting up reservoirs downstream may be of little direct consequence to the people upstream who clear land and cut trees for fuelwood. Downwind as well as downstream effects may be important. In flatter country an individual farmer who cuts his trees will reduce his protection from the evaporative and erosive effects of wind, but the immediate effects on the fertility of his own and his neighbors' soil may be small. The damage becomes apparent only if the weakness of community organizations allows the destruction of trees to spread to many farms or if the trees had enhanced cropping activities under their canopies, as do various species of acacia. In general, both in watersheds and in flatter regions, the social costs of such externalities may be appreciable and are the principal reason that public investment in afforestation programs and in the development and application of forestry laws has become a matter of urgency in Africa. From the third group of explanations noted above it is apparent that such public investments would be fully consistent with—and their success dependent on—policies that are far more favorable to agricultural development than those that have prevailed in the past.

## Notes

1. Consumption rates of families that use fuelwood are generally about 0.6–1.0 cubic meter per capita; one cubic meter is equivalent to about 0.3

tons of oil. See, for example, Nautiyal (1979). A more detailed presentation of evidence on fuelwood consumption as a factor in deforestation is found in Anderson and Fishwick (1984), which contains an extensive bibliography. Two recent reviews that look at land clearing and at other factors as well are Brown and Wolf (1985) and Timberlake (1985), which also contain good bibliographies.

2. The estimates are unavoidably crude because little local measurement of stocks and MAIs is being undertaken. The FAO *Yearbooks* provide some data; for an interpretation see Allen and Barnes (1985). See also Keita (1982) and Clément (1982) and the references in Timberlake (1985) and in Allen and Barnes to studies by Myers.

3. In regions where consumption begins to exceed the MAI the rate of change in the volume of tree stocks can be represented as

$$dS/dt = (a - f)S - C$$

where $S$ denotes stocks at time $t$, $C$ the consumption rate, $a$ the MAI per unit stocks, and $f$ the net rate of felling—the rate of removal of trees for agricultural, industrial, and other activities minus the rate of planting by the forestry service and by farmers. (The value $f$ is not necessarily constant but is here put as a parameter associated with $S$ on the grounds that current planting rates are small and that the rate of land clearing declines with $S$.)

The consumption rate is thought to increase exponentially with population growth. But as fuelwood becomes scarce, real costs and prices rise, and people turn to substitutes or otherwise reduce consumption. Hence we might write the consumption rate as $C = (A/P) \exp(pt)$, where $A$ is a constant, $P$ is price, and $p$ is the population growth rate. In turn the price increases might be expected to vary inversely with (among other things) the available stocks, from which

$$C = (C_0/S_0)S \exp(pt)$$

where $C_0/S_0$ is the ratio of the consumption rate to stocks at $t = 0$. This gives

$$dS/dt = S[a - f - (C_0/S_0) \exp(pt)],$$

the solution to which, taking the case where $f$ is constant, is

$$S/S_0 = \exp\{(a - f)t - (C_0/pS_0)[\exp(pt) - 1]\}.$$

The term inside the exponent includes an exponential term, showing that the *rate* of decline of tree stocks, under the circumstances considered, itself has an exponential element. For example, assume that there is no felling except for fuelwood and that the yields from new plantings are small ($f = 0$); that the initial consumption rate is just equal to the MAI ($C_0/S_0 = a$); that the MAI is equal to 2.5 percent of stocks ($a = 0.025$); and that the population growth rate is 3 percent a year. Then the annual rates of decline of stocks at 0, 10, 20, 30, 40, and 50 years are 0.0, 1.0, 2.1, 3.6, 5.8, and 8.7 percent, respectively, and the stocks decline to 19 percent of their original value over 50 years. If stocks are initially declining at, say, 2 percent a year (compare the data in Allen and Barnes 1985), the decline to 19 percent takes only 30 years, and even quite modest assumptions about $f$ reduce the time to 15 to 20 years. A similar equation can be written for the rate of radial spread of deforestation from fuelwood-deficit areas, with the costs of transport being a function of the radius.

4. In Nigeria fuelwood is currently being trucked 200 kilometers or more to the cities in the north. See also the case study in Tanzania by Allen (1985).

5. See Brown and Wolf (1985) and Lanly (1982). A reviewer of this chapter commented that "the bulk of the literature suggests that the principal cause of removal of forests and trees is clearance of land for agriculture, with overgrazing, and the burning associated with grazing, and fuelwood harvesting being subsidiary causes."

6. In 1986 the Ivorian government told the World Bank that only 2.8 million of the original 16 million hectares of the tropical rain forest remained.

7. For an example of the relative contributions of farm trees, woodlands, and forests to the MAI, see the data in table 2, chapter 3, for Kano State, Nigeria.

8. The literature on the subject is huge. For further reference material and reviews see Allen and Barnes (1985), Anderson and Fishwick (1984), and Timberlake (1985).

9. See also Lamprey (1978) for an assessment of the causes of desert encroachment in Kenya.

10. This area needs more social and anthropological research, modeled on Hill (1972, 1986). The social science bibliography on natural resources in Africa compiled by Seeley (1985), which Megan Vaughan drew to my attention, contains additional references.

CHAPTER 2

# Policy Options

IT IS APPARENT from the preceding chapter that the development
of African agriculture requires stronger measures to maintain soil
fertility and halt desertification. Such measures are necessary
adjuncts to other (perhaps more familiar) policy instruments
intended to raise investment in and output of agriculture—price
policies for crops and the provision of rural infrastructure and
institutional services.[1]

Beyond the need for proper implementation of laws for the restoration, protection, and maintenance of forest reserves, two
complementary forms of public investment in afforestation
should be pursued. We have labeled one the forestry approach, the
other the farm forestry approach. The forestry approach is charac-
terized by plantation-type projects to restore and maintain forest
areas and woodlands and to establish watershed plantings and
shelterbelts for erosion control and for the production of
fuelwood and lumber. In Africa such investments are generally
undertaken by the forestry service. The farm forestry (also called
the agro-, social, or rural forestry) approach involves the planting
and maintenance of trees and shrubs by the farm families themselves near dwellings, as windbreaks, on the edges of cropland, in
copses, and in small lots, sometimes in conjunction with crops.
(Although foresters use the terms *shelterbelt* and *windbreak* inter-
changeably, a useful distinction can be made between multiple
rows of trees extending for several kilometers—shelterbelts—and
shorter, single-row, boundary plantings—windbreaks—which
are what farmers are more likely to undertake. For the purposes of
this analysis, however, the more important distinction is that
between public and private plantings, which have important economic and institutional differences. In this book private plantings
of shelterbelts and windbreaks are considered under farm forestry and public plantings under forestry.)

Foresters and agriculturalists have long recognized the importance of these approaches, which are incorporated in the forestry
laws and public expenditure programs of several countries. But
despite their ecological and economic importance, forestry and

farm forestry investment, maintenance, and management activities have not received significant support in the public expenditure programs of Sub-Saharan African countries. Nor is there evidence that private investment is taking place autonomously on a sufficient scale. Indeed, farm forestry practices, once common in some parts of Africa, are disappearing under the demographic and economic influences noted in chapter 1. This chapter reviews the economic merits of investments in forestry and discusses the main elements of afforestation programs.

## The Forestry Approach

The planting programs of the forestry services have so far accounted for a large share of governmental and external budgetary allocations for investments in fuelwood, erosion control, and related purposes in Africa. Rising scarcity prices for fuelwood in many places indicate that good financial and economic rates of return could be had from, for example, periurban fuelwood plantations in areas where local reserves are badly depleted and from multipurpose plantations to provide fuelwood and poles and control erosion in rural areas. More important, however, shelterbelts and watershed plantings have appreciable ecological and economic benefits. First, consider the benefits of shelterbelts, which may be divided into four components.[2]

1. The prevention of further erosion and thus of losses in soil fertility in the sheltered areas. The economic benefits can be measured by comparing the net present value (NPV) of cropping and livestock activities in the sheltered areas with the NPV of such activities if the areas are left unprotected (in which case yields may be expected to decline).

2. Increases in average crop yields owing to reduced wind velocities and increased soil moisture. The common assumption of a 20 percent net effect on yields (after allowing for the area, moisture, and nutrients taken up by the trees themselves) has been verified by numerous field measurements in countries where shelterbelts have been planted during the past sixty years. Some countries have reported a finding that is potentially significant for the arid and semiarid regions of Africa but that has yet to be confirmed by measurements in these regions: in dry years the effect on yields may be much greater than 20 percent because the marginal benefits of the added soil moisture are greater (see chapter 4). For

example, whereas crop yields may fall by one-half or more in unsheltered areas in dry years, sheltered areas may experience only moderate declines. Such possibilities obviously merit further research.

3. Increased crop residues, and thus more dry-season fodder for livestock, as a consequence of higher yields. For example, a ton of sorghum or millet grain may be associated with up to two or three tons of stover (cured stalks), and thus a 20 percent increase in grain yields can produce half a ton of stover. This source can be further supplemented by plantings of fodder trees and shrubs on the farms in the sheltered areas. Elementary livestock budget exercises show that the gains are significant.

4. The value of the wood itself.

Watershed plantings have similar economic effects. An additional benefit may be alleviation of siltation of downstream hydroelectric and irrigation projects.[3] Figure 1 summarizes the benefits of good watershed management.

This analysis of the benefits of forestry programs brings out the economic effects of associations between agricultural, pastoral, and silvicultural activities. Such associations are immensely important, and to neglect them is to underestimate what might be accomplished through forestry programs. In a study of shelterbelts in the arid zone of northern Nigeria, for example, we found that the rates of return on wood production were less than marginal (around 5 percent) but that these benefits were small in comparison with the first three items, being less than one-fifth of the overall benefit stream under conservative assumptions about the ecological effects. (The results of the study are presented in chapters 3–5.)

It is often argued that the long lead times of forestry investments make certain benefits insignificant because of discounting. Suppose, for example, that the average annual benefit stream of protecting soil fertility (benefit 1) is $B$, once a shelterbelt program has reached maturity, and that the gestation period is fifteen years. Then the present value of benefits at a 10 percent discount rate is only 2.4$B$, which may be small in relation to capital costs. One problem with such calculations is that estimates of $B$ are generally based on the current net value of output in traditional agriculture; that is, they ignore the possibility that agricultural productivity might rise under favorable economic policies or in response to the effect of population pressure on land values. It is, however, inconsistent to discount the returns to such an economically important

Figure 1. *Watershed Management: Physical and Economic Effects*

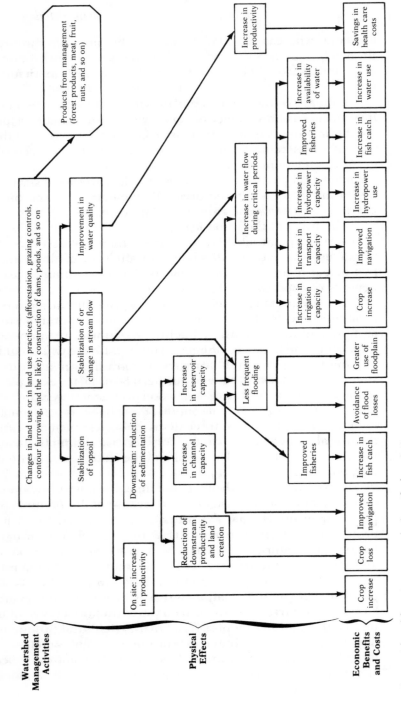

*Source:* Adapted from Gregersen and others (forthcoming).

18

activity as agriculture at 10 percent, and also discount the possibility that its productivity may rise. Since the opportunity cost of capital is significantly influenced by agriculture, we must assume either a lower discount rate and no growth in agriculture or a higher discount rate and growth in agriculture. If the latter assumption is made and we take (say) a 3 percent growth rate of $B$, it is readily shown that the benefits derived from the static assumptions are underestimated by more than a factor of two. Parallel remarks apply to benefits 2 and 3.

Elementary benefit-cost calculations of the kind outlined above suggest that public shelterbelt and watershed plantings have prospects of good returns if at least one of three conditions obtains: (a) the area in question is ecologically threatened but still well farmed, such that benefits 1, 2, and 3 are significant; (b) the downstream costs of siltation of existing or prospective hydroelectric or agricultural projects are appreciable; and (c) a combination of high scarcity prices and high mean annual increments (MAIs) leads to high wood benefits. In subhumid and humid zones, say above 800 millimeters of average annual rainfall, the third condition alone may justify the investment, with the first two serving to emphasize its urgency, but frequently it is the combination of all three that is important. (Much the same conclusion emerges from the 1986 study by French on deforestation in Malawi.) The analysis places much weight, as it should, on afforestation to prevent ecological deterioration in areas still farmed; the returns tend to be low in abandoned areas where the costs of recovering the land for productive purposes are high (see Booth 1979, and chapter 3, below).

Despite the economic importance of the forestry approach, it has one overriding limitation: what it might accomplish falls far short of what is needed to meet rural and urban demands for fuelwood and to protect soils in rural areas from erosion and losses in fertility. With respect to fuelwood, the World Bank's energy policy paper (1983b) concluded, on the basis of simulation studies by John Spears, that even if the demand for wood were reduced by 20 to 30 percent through conservation and substitution of other fuels, a fifteenfold increase in current planting rates would be required in African countries to improve the balance between demand and supply by the year 2000—that is, to arrive at a point where the MAI of wood production is roughly equal to the amount of wood being extracted annually for fuel. These results were averages for quite large regions, and since they ignored the loss of trees owing to land clearing, they must be considered conservative. In regions where fuelwood is scarcer and land clearing is extensive much larger increases in planting rates would be needed. Another way of looking at planting requirements is to recall that plantations currently supply less than 5 percent of the fuelwood consumed; over 95 percent comes from the felling and lopping of trees on natural woodlands and farm fallow.

The budgetary and institutional development problems that have to be surmounted to achieve such an increase in planting rates are formidable and come at a time when forestry services in the region have received appreciable budgetary cuts. Although significant increases in budgetary allocations to the forestry services are desirable, increases of the order required to achieve a fifteenfold increase or more in the planting rates and, equally important, to maintain the plantations and existing forest reserves are not probable. In addition, there are the difficulties of recruiting, training, and managing the labor required for a huge expansion of the forestry service. For these reasons, and because of the ecological and economic importance of the alternative approach, plantings by the forestry service need to be complemented by plantings by villages and farmers—that is, by farm forestry.

## The Farm Forestry Approach

Farm forestry practices complement the programs of the forestry service in the following ways.

1. Higher planting rates. Farmers outnumber foresters, and if their families spent, say, ten days a year on the planting and care of trees as an off-farm activity, their combined labor input would outweigh that of the forestry services. If the farmers are given adequate incentives to adopt the practice, with suitable encouragement and advice from the rural extension services, and are supplied with seedlings, an appreciable increase in planting rates could be achieved.

2. Savings in public expenditures. Wage labor is the largest expenditure item in the forestry service budget, but in farm forestry most labor is provided by the farm families. Another big item in the budgetary costs of public plantings is the fencing required to protect saplings from livestock. Fencing costs for public

shelterbelts amount to about $1,500–$2,000 a kilometer. Since each shelters about 20 hectares of farmland, fencing costs alone may average $70–$100 per hectare sheltered. Farm trees, in contrast, can be protected individually using thorns and other local materials. Although significant expenditures are required to develop research programs, extension services, nursery networks, and so forth, the budgetary costs of farm forestry are relatively low for each hectare of farmland protected—around one-fifth those of public shelterbelts.

3. Reduced farm costs. Although the labor costs of planting and protecting the trees are transferred to the farms, total farm labor costs may be reduced over the long term because of savings in the time that would otherwise be spent in gathering and carrying fuelwood to the farm. Alternatively, if wood or other fuels are now purchased, tree plantings would bring about financial savings.

4. Benefits to crops and livestock. Ecologically the effects are similar to those described above for public shelterbelt and watershed plantings in farming areas. Soil erosion is reduced and nutrient and moisture content improved for much the same reasons. Furthermore, if a high proportion of farm families in the ecologically threatened areas adopts farm forestry—and a high proportion is an essential element—the actual effects on crop yields and livestock may be greater, since the trees will be better distributed over the farmlands.

5. Planting areas and yields. Only small areas of farmlands are taken up by farm forestry practices. It is estimated that in subhumid and humid regions (1,000 or more millimeters of rainfall a year) forty to sixty trees planted on farm boundaries and near dwellings, occupying less than 3–5 percent of arable land on, say, a 2–3 hectare farm, will meet the fuelwood needs of a farm family. (The yields of freestanding trees are about three times the yields of trees in plantations and forests.) In the more arid Sudanian and Sahelian regions yields are significantly lower and self-sufficiency in fuelwood supplies probably cannot be attained. Nevertheless, even where self-sufficiency is not possible, the practice would contribute to wood supplies, provide by-products such as fruit and fodder, reduce erosion, and generally improve the carrying capacity of the soil.

Since the benefits of farm forestry would ultimately be revealed in farm incomes, why do farmers not adopt the practice autonomously? Again, reference must be made to the points raised in chapter 1 concerning land tenure, farmers' perceptions of the problem, the lack of incentives for agriculture and for investment in land, declining or stagnant farm incomes, the harvesting of common woodlands to supplement incomes, and the weaknesses of government and community organizations in working with farmers to address the external costs of deforestation. Whatever weight is given to these and other factors, public agencies face appreciable risks and difficulties as they try to foster farm forestry practices. In a recent afforestation program in the arid zone in Nigeria (discussed in greater detail in chapter 3) attempts to develop farm forestry were almost completely unsuccessful, except for plantings of fruit and shade trees. Seedling survival rates were very low; it is thought that less than 5 percent of over 50 million seedlings distributed free of charge during a five-year period survived. Often the potting soil was used as fertilizer for crops and the seedlings were discarded. Where soils were already badly eroded or where water was in short supply, seedlings did not survive. Because of delivery problems many seedlings were dead on arrival. Others were destroyed by livestock or were planted too late because farmers were busy planting crops.[4] Such difficulties are no doubt symptomatic of the deeper problems of tenure, economic incentives, and institutions noted above; they are cited here only to emphasize the problems in developing farm forestry.

Nevertheless, and despite the risks and difficulties involved, the only financially feasible alternative in sight is to pursue the farm forestry approach to afforestation in conjunction with other programs such as public investment in shelterbelts, watershed plantings, and general forestry management. Much has been learned in Africa, as in Asia and Latin America, from the experiences of past programs, and there is now a copious literature on the design of farm forestry programs for different ecosystems and social systems.[5] The elements required are well known. They were described in my paper with Fishwick (Anderson and Fishwick 1984), with special reference to the arid and semiarid zones of West Africa, as follows.

• *Investment in information.* Improved data is needed on fuelwood consumption, tree stocks, yields, depletion rates, and ecological conditions. Also needed are studies of the sociological, economic, and tenurial factors likely to influence people's response to the programs and their choice of trees.

• *Research, training, and education.* The changing emphasis in the arid and semiarid zones of West Africa toward tree planting more closely associated with and carried out by private homeowners and farmers requires fundamental changes in the priorities of forestry research, university curricula, and technical training. Research and extension linkages in forestry scarcely exist today in most West African countries, and forestry is never included in training-and-visit agricultural extension programs. To integrate tree planting with farming activities, extension workers will need to be educated in farm forestry practices, and supporting research will be required.

The forestry research centers, which have been badly neglected in African countries, merit far more budgetary and developmental support than they have received in the past. In the humid and subhumid zones much can be accomplished with existing species, but in the arid zones research is a necessity. It would be advantageous to establish a West African regional center for agro-silvopastoral research or at least to coordinate research. Although tree planting, usually of exotics, has been successfully carried out in the region since the early 1900s, success was largely attributable to the plantations' being on superior sites and to better rainfalls than have been experienced in the past sixteen years. Recent droughts and the dearth of good soils have underscored the need for improved technological packages and for breeding programs that use survivor trees.

• *Nurseries, seedling distribution, and planting techniques.* A key to fostering "spontaneous" tree planting efforts is the establishment of a highly decentralized nursery and seedling distribution system that is responsive to the farmers' demands.[6] In addition to providing seedlings, public and community nurseries could have a valuable demonstration effect for private investments.

The point about the role of extension services helps to explain why the first attempt at a farm forestry program in northern Nigeria (discussed further in part II) failed; the seedlings were distributed to farmers without supporting research, extension, or monitoring of farmer response and seedling survival rates. Another problem was that the attempt to reach more than 2 million farm families in the region from the outset greatly diluted the impact of the program. In general, the areas and numbers of farms covered by farm forestry programs must be related to the capacities of the supporting research, education, and extension services if the programs are to have a chance of success.

## Conclusions

Several studies have identified workable policies for addressing the ecological and economic problems arising from fuelwood consumption and land clearing in Africa. The importance of creating a favorable policy environment for agricultural investment through economic incentives and the provision of supporting infrastructure and institutional services is self-evident. In addition, there is a clear need for public investments in afforestation. First, increased budgetary support is needed for the traditional work of the forestry services, particularly in the management of the forest reserves and the establishment of erosion control plantations in watersheds and shelterbelts. The value of the wood produced by such plantations, although often significant, is only a small fraction of the benefits. Both watersheds and shelterbelts offer three other benefits that may be substantial in the still well-farmed but ecologically threatened areas and that are not analytically difficult to quantify: prevention of declines in soil fertility; enhancement of the nutrient and moisture content of the soil, which increases crop yields; and, as a result of higher yields, increases in supplies of stover and dry-season browse. In watersheds there may also be the downstream benefits of reducing siltation of reservoirs and irrigation projects.

Second, it would be desirable to take further the current (and somewhat small and isolated) initiatives to develop farm forestry practices in Africa. The benefits are similar to those noted for public watershed and shelterbelt plantations, but their magnitude may depend on how the farmers respond. Given the difficulties of obtaining a good response from the farmers, the financial risks in farm forestry development programs are not small, but the ecological and economic risks of inaction are surely larger, and the programs hold much promise. By involving the farmers themselves in afforestation, farm forestry programs would greatly reduce the budgetary costs per hectare of farmland protected, perhaps by 80 percent or more, and would make possible a significant increase in tree planting rates. Because of its prospective economic and ecological benefits, its low budgetary costs, and its broad scope, farm forestry merits substantially increased public support.

# Notes

1. Two other commonly discussed policy elements reviewed in Anderson and Fishwick (1984) are the substitution of commercial energy for fuelwood and the use of wood-burning stoves to save fuelwood. For recent discussions on the first see World Bank (1983b) and on the second see Foley, Moss, and Timberlake (1984), Manibog (1984), and Hughart (1979).

2. See National Research Council (1984b), annex A, and chapter 4, below, for further discussions of shelterbelts and their effects on yields.

3. For a recent study of watersheds see Easter, Dixon, and Hufschmidt (1986). Dixon and Hufschmidt (1986) and Hufschmidt and others (1983) contain much relevant work on environmental economics.

4. I am particularly grateful to Ali Abibu and Olare Amosum for many discussions and for their insights into these issues. For some statistics on the program see Nigeria, Federal Department of Forestry (1983).

5. See, for example, Food and Agriculture Organization (1985); Tanzania, Ministry of Lands, Natural Resources, and Tourism (1984); Raintree (1984) and other papers in Wiersum (1984); Mnzava (1983); von Maydell (1983); Burley (1982) on the implications of the experiences of India and Kenya; and von Maydell (1979) and other papers in Chandler and Spurgeon (1979). Griffin (1986b) describes a program in Xinjiang Province, China; reports presented by Chinese officials at a recent symposium on windbreaks are found in Brindle, Hintz, and Sturroch (forthcoming). Blair (1986) reported a very high response rate to the social forestry program in India, although mainly (and unexpectedly) for the production of wood for building materials and commercial purposes rather than for fuel.

6. See especially Blair (1986), Raintree (1984), Mnzava (1983), and von Maydell (1979).

# PART II

A Case Study
of Northern Nigeria

CHAPTER 3

# The Current Situation and Approaches to Rural Afforestation

THE CASE STUDY IN THIS PART examines the prospective economic risks and returns on investments in rural afforestation in the arid zone of Nigeria. This zone comprises the northern parts of Sokoto, Kaduna, Kano, Bauchi, and Borno states (see map).[1] The total area is approximately 170,000 square kilometers, and the topography is flat to undulating, with occasional steep granite outcrops. There are two major river systems, the Sokoto-Rima, which drains northwest from Kaduna State into Sokoto State and then turns southwest into the Niger River, and the Jama'are-Hadejia, which drains northeast from Kano and Bauchi states into the Chad Basin.[2] The area has a harsh climate with a long dry season and periodic droughts; annual average rainfall varies from about 800 millimeters over a three-month (but uncertain) rainy season in the south to about 200 millimeters in the northeast, where the rainy season is sometimes less than a month. The estimated population is 20 million, and there is a dense livestock population as well. The destruction of trees in woodlands and on farms because of the demand for fuelwood, agricultural land, and livestock fodder has led to a progressive degradation of the environment and has made the soils more vulnerable to desiccating winds. There is a large fuelwood deficit and a risk of a further decline in the already low productivity of agricultural land.

Many officials and community leaders recognize the severity of the situation, and afforestation in the arid zone is now regarded as having a high economic priority. In 1978 the federal government initiated the Arid Zone Afforestation Program (AZAP). Although by the time the program was curtailed in 1984 it was regarded as unsuccessful, it did provide useful lessons for future investments. This chapter examines the loss of trees in the region and, in the light of the experience gained under AZAP, describes what is cur-

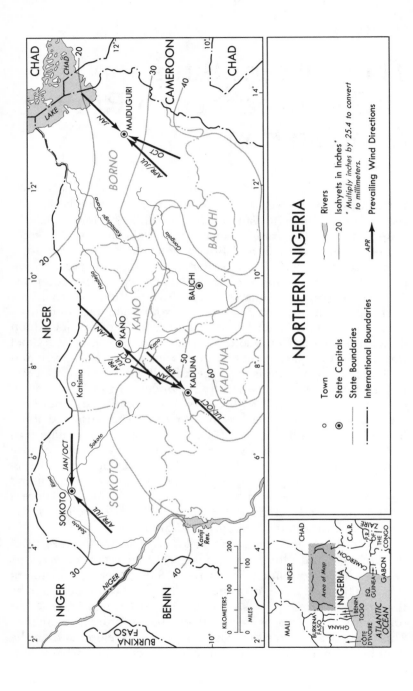

# NORTHERN NIGERIA

Legend:

- o    Town
- ◉    State Capitals
- ------    State Boundaries
- ·—··—··    International Boundaries

- ︵    Rivers
- —20—    Isohyets in Inches*
-    * Multiply inches by 25.4 to convert to millimeters.
- APR→    Prevailing Wind Directions

rently regarded as the most workable approach to afforestation, an approach that is now being pursued by the federal and state governments with the support of the World Bank.

## Losses of Trees

It is difficult to make anything but a very broad quantitative estimate of the rate of tree loss in the region because neither an inventory of stocks nor measurements of their growth (the mean annual increment, or MAI) have been undertaken recently. The main source of information is field reports by foresters; these are not comprehensive, are highly approximate, and rely more on the foresters' judgment and experience than on precise measurements. Hence, one of the first priorities for research and forestry institutions in the region must be to gather relevant empirical evidence on tree stocks, MAIs, consumption rates, soils, and soil erosion rates.

The defects in the information, however, do not obscure the changes taking place in the region. Only the precise magnitudes and locations of the losses in tree stocks are in doubt, not the general trends.

Table 1 shows the areas and estimated yields of the forest reserves of the northern states and gives estimates of the rate of consumption of wood for fuel and for other purposes such as building materials and fencing. The forest reserves account for 9 percent of the area of the northern states and the game reserves for another 3 percent. Fuelwood gathering, subsistence farming, and the collection of cattle fodder by the nomadic Fulani herdsmen encroach on both and cause much degradation of the reserves.

Most of the trees, however, are distributed over the farmlands and fallowlands, which together still account for most of the MAI. This can be seen in table 2. According to the estimates, the MAI of existing trees in Kano State is currently around 0.9 million cubic meters a year (assuming an MAI of 0.2 cubic meters a hectare). Given consumption of around 6.6 million cubic meters a year (see table 1), the net deficit is over 85 percent. Since wastage arising from land clearing is ignored, the above estimates probably understate the deficit. Furthermore, the deficit is likely to be increasing because of population growth, the reduction of the natural regeneration rate (which is proportional to the stocks) as

Table 1. *Forest Reserves and Yields and Wood Consumption Rates in Northern Nigeria*

| State | Area (thousands of square kilometers) | Population (millions, 1980) | Forest reserves (thousands of square kilometers)[a] | MAI of forests (millions of cubic meters a year)[b] | Consumption (millions of cubic meters a year, 1985)[c] |
|---|---|---|---|---|---|
| Sokoto | 95 | 4.3 | 15.3 | 1.5 | 3.2 |
| Kaduna | 70 | 6.2 | 7.5 | 0.8 | 4.7 |
| Kano | 42 | 8.8 | 1.1 | 0.1 | 6.6 |
| Bauchi | 63 | 3.7 | 4.7 | 0.5 | 2.8 |
| Borno | 117 | 4.3 | 7.4 | 0.7 | 3.2 |
| Total | 387 | 27.3 | 36.0 | 3.6 | 20.5 |

a. Excludes game reserves, which account for about 3 percent of the area.

b. The MAI assumes 1.0 cubic meters a hectare a year for forest and woodland reserves. This is thought to be high, since over 80 percent of the reserves are unmanaged and are in the Sahelian and Sudanian belts; some estimates are as low as 0.4 cubic meters a hectare a year (see source note to table 2).

c. The wood consumption estimates assume 0.65 cubic meters per capita a year, of which 90 percent is for fuel and the rest for other uses, and a population increase of 15 percent between 1980 and 1985. The assumption is conservative; for example, the estimates by Nautiyal (1979) are closer to 1.0 cubic meters per capita a year.

*Source:* FAO survey documents prepared during the AZAP project described in the text.

Table 2. *Trees and Their Yields, Kano State, 1980 Estimates*

| Item | Woodland and farmland | Fallow | Forest reserves | Total or average |
|---|---|---|---|---|
| Area (thousands of hectares) | 3,145.0 | 646.0 | 118.0 | 3,909.0 |
| Number of trees per hectare | 3.0 | n.a. | 24.0 | — |
| Volume of wood (cubic meters per hectare) | 5.4[a] | 9.0 | 12.0[a] | 6.2 |
| Total volume (millions of cubic meters) | 17.0 | 5.8 | 1.4 | 24.2 |
| Total volume (percent) | 70.0 | 24.0 | 6.0 | 100.0 |
| Yield (MAI) (cubic meters per hectare a year) | 0.2 | 0.2 | 1.0 | 0.2 |
| Total yield (thousands of cubic meters a year) | 629.0 | 129.0 | 118.0 | 876.0 |
| Total yield (percent) | 72.0 | 15.0 | 13.0 | 100.0 |

— Not applicable.

n.a. Not available.

a. Free-standing trees on farmland are generally mature trees with significantly higher volume per tree than trees in forests.

*Source:* Based on an unpublished field report by Fishwick, 1981. Fishwick, however, judged current yields in forest and wooded reserves to be 0.4 cubic meters per hectare a year; this table and table 1 (from FAO estimates) use 1.0 cubic meters per hectare a year.

stocks decline, and the loss of seedlings and young trees to overgrazing.

The situation is similar, although varying in degree, in the arid zones of the other northern states. It is more severe in parts of Borno and northern Bauchi and is comparable to that in Kano in northern Kaduna and Sokoto. The consequences are fourfold.

• A marked decline in farm tree stocks in the arid zone. Surveys in the Katsina area in northern Kaduna show a reduction in farm trees from 15 per hectare in the mid-1950s to 6.5 per hectare in 1971. This area has a higher density of trees than other locations in the arid zone, and the average for the whole region may not be more than 3 trees per hectare.

• A threatened decline in soil fertility. This has probably occurred already in many areas, although precise measurements are lacking. There are, broadly, five effects, and possibly the start of a sixth: gully erosion, the loss of topsoil to winds, dune formation (already common in the northernmost part of the country), greater surface evaporation and reduced moisture content in the soil as surface wind velocities increase, reduced recycling of soil nutrients, and, possibly, the diversion of sources of soil nutrients (dung and crop residues) for use as fuel if fuelwood becomes scarce.

• The harvesting of tree stocks without replenishment in the southern regions of the northern states for sale in the arid zone. Firewood is transported extensively over distances of 200 to 400 kilometers, and its harvesting is a profitable activity for subsistence farmers and traders to the south. Although this offers the arid zone some respite from the first two effects, it means that the ecological damage is gradually being exported farther south, where tree stocks are still abundant.

• Significant encroachment on and degradation of the forest and game reserves.

In both the arid and the subhumid (southern) regions the loss of farm trees has been further increased by the use of tractors. The transition from the use of hand cultivation and draft cattle to mechanization need not be at the cost of extreme losses of farm trees. Nevertheless, this has been the case so far.

Because of the third effect described above, it is as desirable to encourage farm forestry in the southern as in the northern regions. In the south the investment costs of reafforestation could be avoided because mature trees are already present. Both the protection and the planting of farm trees have been limited in all regions, however, despite governmental attempts to encourage the practice under AZAP. Farmers will take pains to protect some trees—fruit trees, such as cashew, mango, parkia, and guava, and shade trees, such as neem—but these exceptions, although important, do not suffice to offset losses. As discussed below, the required investments in afforestation are appreciable and carry significant risks and difficulties.

## Approaches to Rural Afforestation

Public policies toward rural afforestation in Nigeria currently have two main elements:

• A program of farm forestry that includes the establishment of a nursery network and a seedling distribution system, the use of the agricultural extension service to encourage farm forestry, and support for research.
• A program of shelterbelts.[3]

Both kinds of investment were tried in the 1979–84 afforestation program. Fifty million seedlings were distributed to farmers in an attempt to encourage farm forestry practices, and several hundred kilometers of shelterbelts were planted by the forestry service. In retrospect, and as is commonly acknowledged in Nigeria, the attempt to develop farm forestry was not successful, with the conspicuous exception of the planting of fruit and shade trees in villages. The reasons are discussed in some detail here to make clear the nature of the risks and uncertainties in farm forestry programs and to indicate how the risks might be reduced. (Other policy questions addressed, although briefly, in this section are the use of roadside plantings and the problem of advanced desertification.)

*Shelterbelts*

Shelterbelts planted by the forestry service, although costly, have a good chance of success if the service has the financial resources to undertake plantings. Kano State, for example, has planted over 600 kilometers of shelterbelts since 1968, including 300 kilometers under AZAP, with a success rate (measured in terms of the proportion of the shelterbelts still surviving) of around 70 percent. These programs were, moreover, undertaken in the state's most arid and most threatened areas, with little or no research into, for example, soil conditions and water tables and with no test drillings for hardpans. With the support of a research and site testing unit, a success rate of 80 to 90 percent is considered achievable. The very low success rate in some states—around 20 percent—can be explained more by financial problems and poor choices of site than by technological or economic uncertainties.

Each kilometer of shelterbelt protects 150 to 200 meters of land on the leeward side, depending on tree height. (Plantings of farm trees and hedgerows between the shelterbelts might also help to decrease wind velocity so that the same effect could be achieved with shelterbelts placed farther apart, but this possibility calls for further research.) Hence, 100 kilometers of shelterbelt planted, say, as 20 strips, each 5 kilometers long and 200 meters apart, would protect an area of 20 square kilometers ($20 \times 5 \times 0.2$), or 2,000 hectares. In practice, the shape of the area protected and the length, number, and spacing of the strips depend greatly on the locality. But the principle of concentrated plantings is widely accepted as good practice and is also considered important for achieving a good demonstration effect.

The task ahead is formidable, however. The area that a fairly ambitious but feasible shelterbelt program would protect is small compared with the area that is considered to be ecologically threatened. For example, the cost of planting 2,000 kilometers of shelterbelts over a five-year period to protect up to 0.04 million hectares would be around $10 million. Of the approximately 17 million hectares in the arid zone, at least a third, or 6 million hectares, is in need of afforestation—roughly 150 times the area that could be protected by the shelterbelts. These figures support the conclusion reached in chapter 2: despite the difficulties of developing farm forestry, there is no practical alternative in sight except to pursue that approach in tandem with a shelterbelt program.

*Farm Forestry*

The 50 million seedlings produced by the nurseries of the Forestry Department for the 1979–84 afforestation program were distributed mostly free of charge to farmers and villages throughout the arid zone. The aim was to achieve a regionwide response from the outset, and participation was encouraged by public broadcasting. The agricultural extension services continued to focus on crops and were not involved in the program. Because there was little research and monitoring of the program, it is not known even within broad limits how many seedlings were planted and survived, but the most optimistic guess is less than 20 percent; the actual figure could be around 5 percent. It is said, for example, that many farmers acquired the free seedlings to use the potting soil as fertilizer for their millet. (The potting soils were specially prepared at the nurseries and included manure.) Other farmers made the attempt but lost the seedlings. Reasons for the losses included:

- Insufficient protection from livestock
- Inappropriate choice of species (fruit trees proved quite popular and were protected; these trees were actively promoted later in the program, as were shade trees, which have been planted copiously and with great success even in the driest, most drought-prone villages)

- Soils that were already badly eroded and were losing their fertility
- Tree planting times that coincided with crop planting, so that trees were planted too late
- Failure to water seedlings after planting (many local wells had run dry, and obtaining water for household consumption and for cattle was arduous enough)
- Transport and distribution problems—perhaps as many as half of the seedlings were dead on arrival.

Even a very generous assumption that 40 percent of the seedlings planted survived would give a maximum survival rate of only 20 percent ($0.4 \times 50$).

Aside from such technical and organizational difficulties, the social and economic problems referred to in chapter 1 have undoubtedly been important, particularly the external costs associated with declining tree stocks. To supplement incomes, farmers are harvesting trees for fuelwood over large areas, even at the price of gradually depleting the carrying capacity of the soil. Once the carrying capacity is badly depleted, the practice is to abandon the land permanently and move on to other areas rather than to attempt to restore and maintain the fertility of the soil through adopting farm forestry practices. (In northern Bauchi State, religious and community leaders have expressed concern over the abandonment of land and have recently become more actively involved in tree planting programs.)

Without further research it is not possible to determine the relative weights to be attached to these explanations of difficulties in farm forestry. Each explanation is eminently reasonable in itself, and it is likely that most or all apply to some degree. The fact remains that farmers are abandoning the once traditional practices of farm forestry and are not planting farm trees on a significant scale. A study of the causes merits high priority as a subject for socioeconomic research in the region.

It is important to emphasize, further, that the above list is far from exhaustive and that historical factors, land tenure, and the aftereffects of past economic policies continue to work against the successful establishment of farm forestry practices. For example, the high value of the naira during and after the oil boom, the preference given to industrial investment in public expenditure and incentives policies, and the general shift of public investment to urban areas all did much to undermine agricultural investment and productivity. Insofar as this depressed agricultural incomes, it must also have reduced incentives to invest in the efficient use of land, particularly in noncrop, long-term activities such as tree planting. The problems and risks of developing farm forestry are thus at least as great as those of developing agriculture itself, and the success of the farm forestry programs will be inseparably tied to the success of agriculture in the northern states. Furthermore, the lower are incomes, the more the subsistence farmers will be tempted to harvest and sell (without replenishment) existing tree stocks to augment their incomes and the slower will be the rate of substitution of commercial fuels for fuelwood in local urban areas. (Fuelwood is an economically inferior good, and people shift to commercial fuels as incomes rise.) Finally, there are the problems arising from the conflicts between the growth of nomadic herds, population growth, and the need for settled agriculture.[4]

Beyond the self-evident need to improve the economic and tenurial incentives to invest in agriculture and in the preservation of soil fertility, what are the possibilities for improving the chances of success of afforestation projects, as measured by the participation in the program of a large proportion of farmers and by the achievement of a significant increase in the number of farm trees? Some lessons emerge from the experiences of the 1979–84 afforestation program.

First, the goal of immediate regionwide coverage of about 6 million hectares with perhaps 2 million farm families was unrealistic and undoubtedly diluted the impact of the program. The current emphasis is on obtaining a good farmer response among, say, 5 to 10 percent of the farm families within defined areas. Research and program monitoring (both previously absent) are to be introduced to determine how best to expand the program subsequently. Second, a reorientation of the work and the involvement of the extension services is crucial, as noted in chapter 2. Third, charges for seedlings (and certainly for fruit trees, which have proved to be popular), perhaps sufficient to cover operating costs, would help the program financially and discourage waste. Fourth, the chances of farm forestry practices being adopted are at their lowest in the ecologically most threatened areas, and the most likely outcome is that the land will be abandoned before the trees can have significant ecological effects; in these areas shelterbelts are the only realistic option.

*Roadside Plantings*

Roadside plantings can also help to replenish tree stocks. Significant erosion is taking place along several thousand kilometers of recently constructed trunk roads. In the past roads have been constructed without investing in roadside plantings and erosion control, and there is much evidence to suggest that, apart from damaging local farms, the practice will add greatly to the costs of road maintenance and repairs and shorten the lifetime of the investments. One possibility for financing the costs of roadside plantings and erosion control is to include them in the road investment and to contract the work out to the forestry service or to private firms. The latter approach might also provide a stimulus for the development of private nurseries, of which there are few in Nigeria.

*The Problem of Advanced Desertification*

The costs and difficulties of recovering land for productive purposes in areas where desertification is advanced are exceedingly large, and the chances of success are small. A report by Booth (1979) for the Food and Agriculture Organization concluded that "very severe desertification includes land covered with huge shifting sand dunes, large and numerous gullies, or salt crusts on nearly impermeable soils. Small areas in this condition may be reclaimed at moderate cost; reclamation of large areas will usually be uneconomic except for places where mobile dunes are encroaching on settlements, roads, highways, railways, or high-value irrigated fields." Severe desertification exists mainly in the Sahelian fringes of the region. It is difficult to find an economical alternative for these areas other than to encourage resettlement and to rely on natural regeneration. (Resettlement has, in fact, been proposed to some communities by the state governments but has been resisted, evidently because the terms were not sufficiently advantageous.) The primary aim of the afforestation programs should therefore be to begin the task of preventing losses in the fertility of soils in areas that are still arable. Von Maydell (1979, p. 18) summed up this position well: "Agro-forestry is not a programme aiming at large-scale restoration and regeneration of more-or-less uninhabited semi-desert wasteland; it should preferably be practiced on family or co-operative farms, around villages which, as a rule, have been established on favorable sites."

## Notes

1. The arid zone is sometimes defined as the area north of latitude 12°N, although, to be more precise, the isohyets run diagonally west-northwest to east-southeast.

2. Flows in both systems have been modified by irrigation works which have disrupted traditional irrigated dry-season farming practices on the extensive river flats downstream and have destroyed a considerable area of riverine woodland through siltation and the raising of riverbeds.

3. Shelterbelts or windbreaks could be planted by the farmers themselves, as part of farm forestry, but in Nigeria this practice is very rare.

4. The benefits of settled farming-pastoral activities are increasingly recognized by the Fulani and by farmers, and there is some evidence that the much-needed changes in land tenure systems may be taking place autonomously. The International Livestock Centre for Africa (ILCA) has estimated, for example, that about 30 percent of the Fulani herdsmen had settled permanently by 1979, a further 50 percent practiced limited seasonal migration, and only 20 percent were nomadic. Settlement was also accompanied by a reduction in average herd size, from 80–100 head for nomadic pastoralists, to 30–70 for seasonal and semisettled pastoralists, to 20–30 for settled pastoralists.

CHAPTER 4

# Cost-Benefit Analysis:
# Empirical Background

THIS CHAPTER IDENTIFIES the benefits of afforestation in farming areas and reviews the data used for the estimates of benefits. Given the paucity of empirical research in the arid zone, the intention here is as much to define the kinds of research needed as it is to emerge with a (necessarily approximate) basis for estimating benefits. The costs of afforestation programs in the region are then summarized. For convenience table 9, below, provides a summary of the main parameters and relationships used.

## Components of Benefits

For purposes of analysis it is convenient to divide the benefits of the afforestation program into four components (as in chapter 2). This breakdown of the benefits makes possible a more precise identification of the principal variables and parameters and facilitates sensitivity analysis.

1. The benefits of preventing declines in soil fertility are measured by taking the present value (PV) of all agricultural output from land at present levels of soil fertility and subtracting the PV of output, assuming a decline in soil fertility.
2. Increases in soil fertility as a result of improved moisture retention and nutrient recycling are measured by the PV of the incremental effects of afforestation on crop yield, since farm forestry and shelterbelt programs not only prevent losses in soil fertility but may actually improve fertility.
3. Increases in the output of livestock products (as extra dry-season fodder becomes available from the stover associated with larger crops and from trees and shrubs) are measured by the PV of the incremental livestock production.
4. The value of the tree products—firewood, poles, and fruit—is estimated in the usual way by multiplying the vol-

umes produced by the prices of the products and taking the PV. (Fodder is counted under benefit 3.)

Output from cropping and livestock activities greatly influences benefits 1, 2, and 3. Since it may take seven to fifteen years for trees to have significant ecological effects, it is necessary to consider both the level of farm output and the trends.

Estimates of the financial and economic values of crop output have been made for farms in the arid zone during the preparation of rural development projects. Table 3 summarizes the results for a traditional smallholding of 3 hectares.[1] The gross value of output in cropping activities is about 140 naira per hectare in border prices; the net value of farm output, after deducting the opportunity costs of the owner's labor, is about 15 percent of this figure, or 20 naira per hectare.

Much less information is available on livestock activities, but the use of crop residues for dry-season fodder by the Fulani herds and by the livestock of the settled farmers (including an increasing number of Fulani) is extensive. Neither the number of Fulani nor the number of cattle in the arid region is reliably known because surveys have not been made for many years. Estimates in 1970 showed more than 5 million Fulani with more than 30 million cattle (assuming 50 or more cattle a family) in the northern area.

A rough estimate of the value of livestock activities can be made by considering how many cattle might be supported by crop residues and by fodder from local shrubs and trees. The crop yields per hectare shown in table 3 are associated with roughly 700 kilograms of leaves and 900 kilograms of stalks per hectare;[2] in round numbers each ton of crop is associated with 2.0 to 2.5 tons of residues. Assuming that 50 percent of the stalks are used for fencing and building materials, about 1,200 kilograms per hectare of stover suitable for dry-season fodder would be available, or 1,500 kilograms if an allowance is made for the browse provided by local trees and shrubs. This is sufficient to support during the dry season slightly less than one cow per hectare (a cow equals one tropical livestock unit—TLU—of 250 kilograms) or four 50-kilogram goats.[3]

The simplest model is the rearing of goats. Assuming that three of the four goats that could be supported each year were sold at 2.5 naira a kilogram (in local live-weight farmgate prices), the gross income would be 375 naira per hectare a year, or 112 naira per hectare a year in border prices (using a standard conversion factor—s.c.f.—of 0.3; see note *c* to table 3). This would raise the gross farm income to around 250 naira per hectare a year. Even allowing for the possibility that a large share of crop residues (say, half) might not be used for stover but left to decay as humus,[4] an annual gross farm income of 200 naira per hectare a year seems possible. This is the figure used in the analysis below; net farm income is taken to be 15 percent of gross farm income.

When we turn to trends in farm output, the difficulties of making forecasts are compounded by uncertainty about future crop yields. Agronomic analyses of new seed and fertilizer packages for the region, or even of increased use of fertilizer without new seed varieties, indicate that yield increases of 50 to 100 percent, and commensurate increases in gross and net farm incomes, are possible within five to ten years. Such increases have not been obtained so far, however, and much doubt remains about future yields. Nevertheless, there are several reasons why the possibility of future yield increases—or, more generally, of future increases in farm incomes—should not be dismissed. Even without increased yields, population growth is likely to make agricultural land more valuable. At present, farmers are able to move into as yet unfarmed areas to increase food output in step with population growth, but given the size of Nigeria's present population (nearly 100 million; see World Bank 1987, table 1) and a population growth rate of approximately 3 percent a year, it is likely that the "land frontier" will be reached in ten to twenty years, if not sooner. In these circumstances, economic and demographic factors are likely to lead to one or more of the following changes:

- A rise in real land values
- A rise in yields (perhaps in response to increased land values)
- A rise in food imports.

Either of the first two events would be reflected in an increase in the gross value of farm output in relation to costs and would thus justify applying a trend term (say, 3 percent a year, in line with population growth) to the gross and net values of farm output once the land frontier is reached.[5] The third event might dampen the growth of land values (or the need for yield increases), but the effects might be limited. Any large (nonmarginal) and growing deficit in farm output in relation to the demand for food would lead to increases in food imports, a downward pressure on the exchange rate, and hence a rise in the price of tradables (such as

Table 3. *Arid-Zone Crop Yields and Values, Traditional Agriculture, 3-Hectare Farm*

| Crop | Area planted (hectares)[a] | Yield (kilograms per hectare) | Total yield (kilograms) | Prices (1985 naira a ton)[b] | | Value of crop (1985 naira)[b] | |
|---|---|---|---|---|---|---|---|
| | | | | Border | Financial | Border | Financial |
| Sorghum | 0.9 | 550 | 495 | 197 | 650 | 98 | 322 |
| Millet | 2.8 | 350 | 980 | 197 | 700 | 193 | 686 |
| Cowpeas | 1.4 | 120 | 168 | 400[c] | 1,200 | 67 | 202 |
| Groundnuts | 0.3 | 350 | 105 | 500 | 1,100 | 52 | 115 |
| Total value (3 hectares) | — | — | — | — | — | 410 | 1,325 |
| Total value per hectare | — | — | — | — | — | 137 | 442 |

— Not applicable.

a. Adds up to more than 3 hectares because some areas are double planted.

b. Economic values are expressed in border prices. The border prices in naira assume that $1.0 is equal to 1 naira, with world dollar prices being used for sorghum, millet, and groundnuts. The naira was devalued in 1986 to about 0.25 naira to $1.0, but since the analysis in accounting prices and the conversion factors used anticipated the extent of the devaluation that was needed and that in fact took place one year after the analysis was completed, it was unnecessary to revise the calculations. (The s.c.f. used in the analysis was 0.35 in 1985 prices—1985 being the reference year of the analysis—which works out to roughly 0.25 in 1986 prices.)

c. Assumes a conversion factor of 0.3 (in 1985 prices), since the marginal products of labor and other local inputs in the region are largely determined by the marginal product of traditional agriculture. The figure 0.3 is the ratio of the value of crop output in border prices to its value in financial prices, or 137 ÷ 442 in the table.

*Sources:* Local surveys undertaken during the preparation of rural development projects; border prices for tradable crops (sorghum, millet, and groundnuts) in dollars are obtained from extrapolation of the figures in World Bank (1983a).

food crops) in relation to the price of nontradables (such as unskilled labor, the principal cost element in farm budgets). Again, the gross and net values of farm output would be likely to rise over time.

It is therefore appropriate to make an allowance for an eventual rise in the real value of farm output in the areas that would be protected by the program. Such an allowance reflects the increasing economic importance of protecting farmlands from erosion and losses in soil fertility, given the reality of population growth in the region. In the calculations in chapter 5 a 3 percent annual growth rate is applied to gross and net incomes, beginning in fifteen years (the time when the land frontier might be reached).

We now have the background needed to proceed with the evaluation of the benefits of afforestation programs.

## Prevention of Declines in Soil Fertility

The prevention of declines in soil fertility is by far the most difficult and uncertain benefit to quantify. The main reason is the lack of relevant data on soil fertility and on the extent of erosion where tree stocks are depleted in the region. In addition, the process by which soil fertility is thought to decline is difficult to model, since it involves the formulation of environmental, demographic, and economic relations over regions and over time. For example, declines in tree densities over large areas, as well as the loss of local tree stocks, may be important because of the effect of tree densities on surface wind velocities and microclimates. Also, as discussed in chapter 3, the decline in trees in some areas (as in the arid zone) may be stemmed for a time by the transport of wood from areas where trees are abundant. This gives some respite to the fuelwood-deficit areas but contributes to the spread of deforestation, with negative but quantitatively unknown effects on soil erosion and fertility over larger areas. Another difficulty is that as fuelwood becomes scarce rural families may turn increasingly to crop residues and dung for fuel, depending on farm incomes, and soil fertility may decline precipitously.[6]

Given such uncertainties and analytical difficulties, the only practical course for the analysis of benefits and costs is, first, to try out a range of assumptions that agronomists and others knowledgeable about the region consider reasonable and defensible and, second, to identify the research that is needed to support a more precise analysis. In chapter 5 rates of decline of soil fertility of 0 to 2 percent a year are examined. These rates are applied to the gross value of farm output but not to costs; costs may even increase over time if it becomes harder to work the land. In the with-project case the decline is gradually stemmed and soil fertility is enhanced as the afforestation programs begin to take effect (after about seven to ten years for shelterbelts and seven to fifteen years for farm forestry). In the without-project calculations the decline continues until net farm income becomes zero or negative, at which point the land is taken out of production.[7]

*Increases in Crop Output*

Field experiments show that afforestation programs, in addition to reducing or preventing soil erosion by reducing surface wind velocities, have positive effects on crop yields: surface evaporation is decreased and thus soil moisture is increased, crop damage from storms is reduced, there are slight increases in local humidity (which reduces stress on plants), and there are generally favorable effects on the nutrient content of the soil. The last effect is thought to be more pronounced for farm forestry than for shelterbelts, since several species of farm trees (acacia species are the most commonly discussed) are slow-growing but highly advantageous for nutrient recycling and do not compete with crops planted under their canopies (National Research Council 1984a). Shelterbelts, in contrast, although they produce green mulch, use faster-growing species (such as eucalyptus and neem) and compete for moisture and nutrients with crops on their boundaries. In general, shelterbelts are much more effective than farm forestry in reducing surface wind velocity (except, of course, when farmers themselves plant shelterbelts or windbreaks) but are less effective than farm forestry in recycling nutrients.

Most of the quantitative data regarding crop output refer to shelterbelts. Table 4 shows some measurements of surface wind velocities, crop heights, and crop yields taken at a windbreak in Niger and at a nearly unsheltered control village. The wind velocity and yield measurements follow the same pattern as is found in other studies, with the effects on yields near the midpoint of the windbreak being roughly twice the average effect. Soil moisture content, also measured in the above study, was found to be higher in the windbreak area than in the control area during periods with a high water supply but lower during periods with low water supply, proportionally more moisture having gone into plant growth (see the effects on average plant height in the last row of table 4).

The possibility that shelterbelts may have negative effects on yields is often discussed by foresters and agronomists. Aside from the obvious problem that if the shelterbelts are incorrectly oriented air currents may be accelerated ("funneling"), jetting and turbulence may occur around the tree trunks at low heights, again leading to local acceleration of the air currents. Shelterbelts also compete for moisture and nutrients with the crops on their boundaries. Jetting can be reduced by planting shrubs and hedgerows, which may increase the effects on yields by 5 percent or more over the figures reported in table 4. Competition of shelterbelts with crops is allowed for in the yield measurements and is, in any case, serious only on the boundaries; the figures quoted and used in this book have allowed for this effect. Some more complex negative effects may arise from the entrapment of higher-temperature air (a boundary layer) within the shelterbelt, but if this is significant it seems to be more than offset by the positive effects, and it can be avoided by allowing some air to flow through the shelterbelts. (The aim of shelterbelts is to diminish wind velocities, not to eliminate air flows altogether or to cause turbulence or the entrapment of hot air.) In conclusion, what these possible negative effects point to is the need for care and attention in design and for the avoidance of obvious errors. (See Brindle, Hintz, and Sturrock forthcoming for methods of designing windbreaks.)

The only research studies undertaken in the arid zone were by a government forestry research unit at Kano. The results were inconclusive but indicated positive effects on yields, ranging from 20 to over 50 percent. These are consistent with the results in several other countries. Table 5 summarizes the data collected in fourteen countries over the past sixty years. For the industrial countries (excluding the U.S.S.R.) the average measured effect on crop yields was 22 percent if measurements of +50 percent or more are excluded, 27 percent if measurements of +100 percent are excluded, and over 40–50 percent if all results are included. An almost identical pattern applies to the data from developing countries.

A finding of potentially great importance for the arid zone has been reported consistently in more than a dozen studies undertaken in the U.S.S.R.: the effects of shelterbelts on crop yields are

Table 4. *Effects of a Double-Row Windbreak on Wind Velocities and Crop Yields, Majjia Valley, Niger*

| Item | Control village | Distance from leeward edge of windbreak (tree height = x) | | | | | |
|---|---|---|---|---|---|---|---|
| | | 1.0x | 5.0x | 8.5x | 12.0x | 16.0x | Average |
| Relative wind velocity at height of | | | | | | | |
| 1 meter | 100 | 78 | 47 | 60 | 72 | 63 | — |
| 2.5 meters | 100 | 38 | 46 | 58 | 66 | 67 | — |
| Relative yield per hectare | 100 | 130 | 156 | 122 | 125 | 111 | 129[a] |
| Mean height of millet (meters) | 2.5 | n.a. | n.a. | n.a. | n.a. | n.a. | 3.3 |

— Not calculated.
n.a. Not available.
a. When adjusted for the area used by the windbreak itself, the net increase in average yield is 23 percent.
*Source:* Bognetteau-Verlinden (1980).

far greater in drought years than in wet years. The table below shows some typical results for the U.S.S.R. (see the source note to table 5).

| | Reported increase in crop yields (percent) | Actual yields (kilograms per hectare) | |
|---|---|---|---|
| | | Sheltered | Open |
| Dry years | 284 | 960 | 250 |
| Wet years | 19 | 1,870 | 1,570 |

A greater effect is to be expected in dry years and in droughts because the incremental benefits of additional nutrients and soil moisture are likely to be greater under these conditions. But the magnitude of the reported effect is remarkable and merits investigation under arid conditions. It is interesting that local farmers in northern Nigeria have made similar claims. In an unpublished report to the World Bank on a field trip in 1980 Fishwick reported that "farmers interviewed in the Unguwar shelterbelt area claimed that in the severe drought years 1973–74, when the local shelterbelt had reached an average height of 8 meters, surrounding villages without shelterbelts experienced crop failures of up to 100 percent, whereas Unguwar Magaji had normal if not better crops."

Given the above evidence and the assessments of numerous foresters and agronomists, it seems reasonable to expect a 15–25 percent increment in crop yields in shelterbelt areas as the trees approach mature heights. This is the range of assumptions used below. It is probably conservative for arid conditions, but further analysis is required to determine whether taking a higher range would be justified.

For farm forestry it is more difficult to decide on an appropriate range of assumptions because of the scarcity of data. The nutrient recycling effects are reportedly large in the vicinity of acacia trees (National Research Council 1984a), but again the quantitative effects of farm tree plantings on moisture and nutrients in the soil have not been measured, at least in farming areas in the arid zone. The effect of plantings on surface wind velocity and soil moisture depends on farm tree densities over large areas and thus on farmer response. If only a few farmers plant trees in a denuded area, the benefits from reduced surface wind velocities may be negligible (unless the plantings include windbreaks), but if a large proportion plant fifteen to twenty trees per hectare, the benefits may be significant. The analysis in chapter 5 considers a 5 to 10

Table 5. *The Effects of Shelterbelts on Crop Yields*
(number of studies)

| Economy | Total | Minus 20– minus 10 | Minus 10– 0 | 0–10 | 10– 20 | 20– 30 | 30– 50 | 50– 100 | 100+ |
|---|---|---|---|---|---|---|---|---|---|
| *Industrial* | | | | | | | | | |
| Canada | 3 | — | — | 1 | — | — | 1 | 1 | — |
| Denmark | 30 | — | — | 4 | 12 | 5 | 6 | 1 | 2 |
| Germany, Fed. Rep. of | 12 | — | — | 3 | 4 | 3 | 2 | — | — |
| Italy | 4 | — | — | — | 1 | 2 | — | 1 | — |
| Japan | 1 | — | — | — | — | 1 | — | — | — |
| Romania | 6 | — | — | — | 1 | 1 | 1 | — | 3 |
| United States | 16 | — | — | 2 | 2 | 2 | 4 | 4 | 2 |
| U.S.S.R. | 84 | 2 | 2 | — | 14 | 12 | 7 | 13 | 21 |
| *Developing* | | | | | | | | | |
| China | 5 | — | — | — | 1 | 3 | 1 | — | — |
| Egypt | 5 | — | — | 1 | 3 | — | 1 | 1 | — |
| India | 2 | — | — | — | — | 1 | — | — | 1 |
| Niger | 1 | — | — | — | — | 1 | — | — | — |
| Taiwan | 8 | — | — | 1 | 2 | 2 | 1 | 1 | 1 |
| Turkey | 1 | — | — | — | — | 1 | — | — | — |
| Total including U.S.S.R. | 174 | 2 | 2 | 21 | 40 | 32 | 24 | 22 | 30 |
| Total excluding U.S.S.R. | 90 | — | — | 12 | 26 | 20 | 17 | 9 | 9 |
| Total developing economies | 22 | — | — | 2 | 6 | 7 | 3 | 2 | 2 |

— No instances found.

*Sources:* For Niger, Bognetteau-Verlinden (1980); for other countries, literature search by Magrath for the World Bank, 1979 (bibliography available on request from the Agriculture and Rural Development Department, World Bank). Magrath did not evaluate the quality or reliability of the measurements. See also Brindle, Hintz, and Sturrock (forthcoming) for additional results that generally support those in the table.

percent range of effects on crop output to provide a rough (and perhaps minimal) indication of what might be achieved if farmers respond well to the program.

## Increased Livestock Production

As discussed above, crop residues, farm trees, and shrubs provide dry-season fodder for livestock. The amount of stover available rises linearly with crop yields (see note 2); for example, a 20 percent increase in yields in traditional agriculture produces approximately 350 additional kilograms of stover per hectare. The planting of farm trees and shrubs could certainly raise this figure to more than 500 kilograms per hectare, sufficient to support over one-quarter of a TLU (the equivalent of a goat), and would justify applying the crop yield increments discussed in the preceding section to the yields of livestock.

## Wood and Fruit Production

The direct outputs from shelterbelts and farm trees differ greatly because mainly fast-growing trees (neem and eucalyptus) are used for shelterbelts, whereas the farm trees include a wider variety of species to meet other needs. The main products discussed here are firewood, poles, and fruit; the fourth product, fodder, has been discussed in connection with livestock activities.

FIREWOOD. Firewood prices in the arid zone range from 15 to 20 naira a cubic meter in the rural areas and from 20 to 40 naira a cubic meter in the cities. Prices are rising, although they vary greatly with location. (A recent World Bank–United Nations Development Programme energy assessment report found prices of 50 to 70 naira a cubic meter in the northern states in 1982.) Urban demand is increasingly being met by wood transported 200 or more kilometers from the southern regions of the states. As firewood becomes scarce, it is inevitable that the rural areas of the arid zone will also depend on this source as long as firewood remains the preferred fuel for cooking. Prices in the rural areas could therefore rise appreciably, with a limit being set by the costs of commercial cooking fuels such as kerosene.

It is surprising that substitution of kerosene for wood is not proceeding more rapidly. Kerosene is still highly subsidized in

Nigeria, but even in border prices it is still much cheaper than fuelwood for cooking, especially in the north. The border price of kerosene is about 300 naira a ton, including domestic transport, but 0.17 tons of kerosene is equivalent to one cubic meter of wood and is two to three (say, 2.5) times more efficient a fuel. Thus even if kerosene were not subsidized, the equivalent of one cubic meter of fuelwood would cost only 20 naira or less (300 naira × 0.17/2.5). Adding the costs of appliances would increase the amount to perhaps 25 naira for the energy equivalent of a cubic meter of wood. The main explanations for why fuelwood is still preferred in places where it costs more than kerosene are (a) retail margins for low-volume distribution of kerosene that are higher than generally thought, and (b) custom. Allowing for (a) and assuming that (b) will change as fuelwood becomes scarce, the economic value of fuelwood is taken to be 30 naira a cubic meter; that is, it is assumed that kerosene prices will put a ceiling on the rapid growth of fuelwood prices and will lead to more rapid substitution than has so far occurred in the northern states. From this are deducted the costs (mostly labor) of cutting, hauling, and chopping, so that the prices are referred back to the stump (unharvested) condition. These costs are about 15 naira a cubic meter in financial terms, or 5 naira when multiplied by the standard conversion factor (s.c.f.) for local labor, giving a net economic value of 25 naira a cubic meter (30 − 5).

POLES. The financial value of the poles is currently about 50 naira a cubic meter; it is assumed to stay roughly constant in line with the costs of production elsewhere in the country. The economic value is obtained by multiplying the financial value by a national average s.c.f. of 0.35 (further discussed in "Costs," below) and deducting the costs of local labor (3 naira in border prices) used in felling, extracting, and loading at the shelterbelts. This gives a net value of about 15 naira a cubic meter, again at the stump, or unharvested, condition.

FRUIT. Market prices in 1985 in the arid zone are as follows in naira a thousand. (Data are from interviews with local vendors.)

|         | Actual prices | Border prices |
|---------|---------------|---------------|
| Mangos  | 65            | 20            |
| Cashews | 33            | 10            |

A local conversion factor (l.c.f.) of 0.3 was used to convert actual prices into border prices (see note *c* to table 3). Given the loss of fruit to animals and pests, waste, and so on, and the costs of gathering and marketing, the net value may be only about 10 to 15 percent of the gross values given above.

WOOD AND FRUIT BENEFITS PER HECTARE FARMED. The costs and benefits are calculated below in terms of per hectare farmed, since this is analytically more convenient and displays the effects of ecological change more clearly. For *shelterbelts* average yields are about 9 cubic meters of fuelwood and 12 cubic meters of poles per kilometer, or approximately 7 cubic meters of wood per hectare of land occupied by the shelterbelt (which takes up 2.8 hectares per kilometer). Using the above border prices, the economic value of wood produced is 432 naira per kilometer of shelterbelt once the trees have reached maturity, or 22 naira per hectare of farmland protected (since each kilometer of shelterbelt protects 20 hectares of farmland).

For farm forestry table 6 shows typical farm trees that might be planted under the afforestation program. An allowance for low survival rates has to be built into any program; for example, if the aim is to achieve twenty surviving trees per hectare, as many as a hundred seedlings per hectare may have to be distributed eventually. Application of the preceding price and cost data to the production figures shown in the last column of table 6 gives a net benefit of 7 naira per hectare a year in border prices. It should be added that the fixed costs of establishing a nursery network and providing regular extension advice to a given number of farmers are large, but the marginal costs of supplying extra seedlings are small. Thus if it were decided to increase the number of farm trees per hectare to get a more pronounced ecological effect, one could distribute nearly twice as many seedlings from the same nursery network to the same farmers with comparatively little effect on costs.

## Production Schedules: The Development of Benefits over Time

The above estimates of the components of costs and benefits all relate to the time when shelterbelts and farm trees are mature and have their full ecological effects. These effects, of course, appear

Table 6. *Farm Forestry: Seedling Survival Rates and Mean Annual Yields of Trees*

| Type | Number of seedlings per hectare | | Survival rate (percent) | Yield per hectare |
| | Distributed | Surviving | | |
| --- | --- | --- | --- | --- |
| Total | 100 | 20 | 20 | — |
| Fuelwood | 20 | 3 | 15 | 0.1 cubic meter of fuelwood a year, all trees |
| Fuelwood and fodder | 50 | 10 | 20 | |
| Mango | 15 | 3–4 | 25 | 1,000 (fruit) |
| Cashew | 15 | 3–4 | 25 | 1,000 (fruit) |

— Not applicable.
*Source:* Foresters' calculations.

only gradually, a number of years (depending on species) after planting.

Both neem and eucalyptus have been used successfully for shelterbelts in the arid zone. The first harvest is usually at year 10 for neem and at year 8 for eucalyptus. For the cost-benefit calculations the following buildup of benefits, derived from foresters' estimates, has been assumed. (The figures given are proportions of the benefits at maturity of the trees. The yields of wood after year 10 have been averaged over time.)

| | | | *Year* | | |
|---|---|---|---|---|---|
| | 1–6 | 7 | 8 | 9 | 10+ |
| *Benefit* | | | | | |
| Ecological | 0.0 | 0.2 | 0.5 | 0.7 | 1.0 |
| Wood | 0.0 | 0.0 | 0.0 | 0.0 | 1.0 |

For farm forestry the benefits similarly build up steadily as the trees approach maturity. Wood and fodder trees usually begin to give the yields noted above about eight years after planting; mangos and cashews can yield fruit eight and six years, respectively, after planting, although not at the rate of a fully mature tree. The following buildup of benefits, derived from foresters' estimates, is assumed. (The figures given are proportions of the benefits at maturity of the trees. The yields of wood and fruit after year 15 have been averaged over time.)

| | | | | | | *Year* | | | | | |
|---|---|---|---|---|---|---|---|---|---|---|---|
| | 1–5 | 6 | 7 | 8 | 9 | 10 | 11 | 12 | 13 | 14 | 15+ |
| *Benefit* | | | | | | | | | | | |
| Ecological | 0.0 | 0.0 | 0.0 | 0.1 | 0.2 | 0.3 | 0.4 | 0.5 | 0.7 | 1.0 | 1.0 |
| Wood and fruit | 0.0 | 0.1 | 0.2 | 0.3 | 0.4 | 0.5 | 0.6 | 0.8 | 1.0 | 1.0 | 1.0 |

## Costs

Financial costs are converted into economic costs in the standard way (a) by estimating the tradable components in border prices directly (May 1985 figures at 1.0 naira = $1.0); and (b) by converting the nontradables into border prices using conversion factors. For manual labor (the principal nontradable input into the project) a local conversion factor of 0.3 is used, based on the value of the output of traditional agriculture in border prices. For all other nontradables (mainly poles and local professional labor), a standard conversion factor (estimated at 0.35) is used; this was the estimated ratio of the shadow to the official exchange rate in May 1985.[8]

Aside from such adjustments in prices to allow for distortions in the exchange rate and for other factors, there are two other ways in which the economic analysis of afforestation differs from the financial analysis: the costs of farm labor are included in the economic analysis of farm forestry, and the land occupied by shelterbelts is costed at its economic value rather than in terms of the financial compensation paid to farmers by local governments. The economic value of the land is the present worth of the net farm income from the cropping and livestock activities that would otherwise be conducted on the land occupied by the trees; it is estimated in the without-project scenario.

*Shelterbelts*

Table 7 shows recent estimates of the financial and economic costs per kilometer of shelterbelts. The economic costs (excluding land) come to 120 naira per hectare, since each kilometer of shelterbelt protects 20 hectares. An allowance has to be made, however, for the possibility that some shelterbelts will fail because of unfavorable water tables, "pans," droughts, encroachment of cattle, and so on. Kano and Sokoto states have had about 70 percent success rates in the past, even without adequate research and testing facilities; with improved facilities an 80 percent success rate is considered achievable. Assuming the latter figure, the economic costs (excluding land) come to 150 naira per hectare.

Table 7. *Shelterbelts: Costs*
(naira a kilometer)

| Item | Financial costs | Economic costs |
|---|---|---|
| Land | 1,420 | Net farm income on area occupied by trees[a] |
| Fencing | 1,000 | 1,000 |
| Posts | 2,040 | 715 |
| Operations[b] | 1,950 | 585 |
| Total | 6,410 | 2,300, plus Land, above |
| Total including contingencies (4 percent) | 6,706 | 2,392, plus Land, above |

a. Estimated in the without-project case on the basis of the value of the land taken out of production. For the method of estimation see chapter 5 and the appendix.

b. Includes seeds, fertilizers, boron, and labor.

*Source:* Project studies in the region.

*Farm Forestry*

The costs of farm forestry include the establishment of nurseries; the training and use of agricultural extension agents to disseminate the ideas and practices of farm forestry; various management, research, and monitoring costs; and the costs to the farmers. The costs shown in table 8 are based on a network of approximately thirty nurseries, each capable of producing about 250,000 seedlings a year for delivery to local villages and farms, and on the costs of distributing the seedlings and providing extension services for about 10 percent of the farm families (250,000 families) in the arid zone.[9] The farmers' costs assume about fifteen days' work to plant and water 100 seedlings and provide protection (local thorns, baskets, mud walls, and so on) for them and a local labor cost of 1.5 naira a day in border prices. This gives a cost of 40 naira per hectare, of which nearly 40 percent (the largest cost item in the program) would be borne by the farm families themselves.

**Summary**

Table 9 summarizes the main variables and parameters used in the analysis of the effect of afforestation on traditional agriculture in the arid zone. With the project, gross and net farm incomes from cropping and livestock activities are lower in the early years than without the project because land is taken out of production by the shelterbelts and (to a much smaller extent) by farm tree plantings (row 3 of table 9; see also figures 2 and 3 in chapter 5).[10] As the

Table 8. *Farm Forestry: Costs*

| Component | Millions of naira | Percent | Naira per hectare farmed |
|---|---|---|---|
| Cost to project | | | |
| Nurseries | 5.8 | 33 | 13 |
| Training and extension | 1.4 | 8 | 3 |
| State management | 3.0 | 17 | 7 |
| Project management | 0.8 | 5 | 2 |
| Subtotal | 11.0 | 63 | 25 |
| Cost to farmers | 6.5 | 37 | 15 |
| Total | 17.5 | 100 | 40 |

*Source:* Investment studies in the arid zone.

Table 9. *Variables and Parameters for Benefit-Cost Analysis*

| Quantity | Value | Remarks |
|---|---|---|
| 1. Gross farm income ($Y_t$) (year $t = 0$) | 200 naira a hectare | See table 3 and accompanying text |
| 2. Net farm income | $Y_t - 0.85\ Y_0$ | Rises in later years (see row 9) |
| 3. Area occupied by trees | | |
| Shelterbelts | 12 percent | Zero without project |
| Farm forestry | 2 percent | Zero without project |
| 4. Rise in gross farm income owing to ecological effects | | |
| Shelterbelts | 15–25 percent | Case with no rise in income also studied |
| Farm forestry | 5–10 percent | Case with no rise in income also studied |
| 5. Year in which ecological effects appear | | |
| Shelterbelts | Year 7–10 | Gradual rise in gross farm income from 0 to 15–25 percent |
| Farm forestry | Year 8–13 | Gradual rise in gross farm income from 0 to 5–10 percent |

| | | Range of values studied |
|---|---|---|
| 6. Rate of decline of soil fertility | 0–2 percent | |
| 7. Year in which decline in soil fertility ceases | | |
|    With project | Year 8 | |
|    Without project | Indefinite | Land abandoned when net farm income is negative |
| 8. Net value of wood | | |
|    Shelterbelts | 22 naira per hectare | Zero without project |
|    Farm forestry | 7 naira per hectare | Zero without project. Includes value of fruit |
| 9. Agricultural productivity | | |
|    Growth rate ($g$) | 3 percent | Applied to gross and net farm income |
|    Year in which $g$ begins | Year 15 | Traditional agriculture until year 15 |
| 10. Project costs (referred to year 1)[a] | | |
|    Shelterbelts | 150 naira per hectare | Range of 145–165 naira per hectare studied |
|    Farm forestry | 40 naira per hectare | Case of 60 naira per hectare studied |

a. Costs assume an 80 percent survival rate (20 percent failure rate) for shelterbelts and survival rates of 25 percent for farm forestry fruit trees, 15 percent for fuelwood trees, and 20 percent for fuelwood-fodder trees.

trees reach mature heights, however, favorable ecological effects begin to appear; the declines in soil fertility are first arrested (row 7) and then reversed (rows 4 and 5) as the output-enhancing effects of the project take hold. There are also wood and fruit benefits (row 8). Without the project, in contrast, the declines in soil fertility continue until net farm income becomes negative and the land is abandoned (row 7). Of importance to the region is the prospective growth in long-term agricultural productivity (row 9), since the better the prospects for productivity growth, the more important it becomes to protect the fertility of the soil. Row 10 shows project costs, which for farm forestry also include the costs to the farmers of planting and protecting the trees.

## Notes

1. The economic analysis is in shadow, or accounting, prices. For tradable goods these prices are usually assumed to be equal to the border value—the cost, insurance, and freight (c.i.f.) or the free on board (f.o.b.) price expressed in units of domestic currency at the official exchange rate. The domestic values of nontraded commodities are converted into border values with the use of conversion factors (see Ray 1984). Border prices are usually different from actual prices (which are used in financial analysis) because of distortions in the exchange rate and, often, in domestic prices. At the time of this analysis the major distortion in the Nigerian economy was the highly overvalued exchange rate; since the accounting prices used allowed for this distortion the analysis did not need to be revised after the devaluation of the naira in 1986. Another way of looking at the problem is to say that since the analysis is valid in 1985 border prices at $1.0 = 1.0$ naira, it is also valid in 1985 world dollar prices.

2. The regression relationships are given in Powell (1984). For the crop mix shown in table 3 they average around $150 + 2.6x$ for sorghum, $260 + 4.2x$ for millet, and $300 + 0.8x$ for groundnuts, where $x$ is the crop yield in kilograms per hectare. Because $x$ is the yield, an allowance for the effect of rainfall on yield is included in the equations, although Powell cautions that the equations were not tested on dry-year data.

3. Le Houerou (1980) estimates that a Sahelian TLU of 250 kilograms must ingest about 1,200 kilograms of stover and 500 kilograms of browse during the nine-month season.

4. It is already assumed that 50 percent of the stalks are used for building materials.

5. It is interesting to read Marshall's comments on population growth and change in land values, written in 1890:

Independently of any change in the suitability of the prevailing crops and methods of cultivation for special soils, there is a constant tendency towards equality in the value of different soils. In the absence of any special cause to the contrary, the growth of population and wealth will make the poorer soils gain on the richer. Land that was at one time neglected is made by much labor to raise rich crops . . . Leroy Beaulieu has collected several facts illustrating this tendency of poor lands to rise in value relatively to rich. He quotes the following figures, showing the rental in francs per hectare of five classes of land in several communes of the Départements de l'Eure et de l'Oise in 1829 and 1852 respectively:

| Year | Class I | Class II | Class III | Class IV | Class V |
|------|---------|----------|-----------|----------|---------|
| 1829 | 58 | 48 | 34 | 20 | 8 |
| 1852 | 80 | 78 | 60 | 50 | 40 |

(The quotation and table are from Marshall 1890/1980, p. 135 and note.)

I was unable to compile comparable data for Nigeria, although there are many reports that land values are already not low and are rising in the northern regions. According to the recent reviews on the rephasing and recosting of the rural development programs, the cost of land clearing in the northern states is more than 500 naira per hectare; land rents in much of the project area are 60 to 120 naira per hectare a year, depending on the fertility of the soil. There is much resistance by farmers to providing land for shelterbelts unless full compensation is paid, at 650 naira per hectare plus (on average) 50 naira per tree, and, there is widespread intrusion in and settlement on the forest reserves.

6. Henrik Hvidberg-Hansen commented in a personal communication that if prices rise sufficiently while incomes remain low, rural people will often burn dung and crop residues, which of course accelerates the rate of reduction in soil fertility.

7. When land is taken out of production, there are the additional costs of migration and resettlement of the farmers and, probably, the costs of further land clearing. These are not allowed for in the following analysis, although they may be significant (as David Pearce has commented in a personal communication). In addition, land that is returning a negative net income (if farm labor is valued at its opportunity cost, as in the farm budget models underlying table 3) may be farmed until the farmers realize that they would be better off if they abandoned the land. In other words, there are costs associated with variances as well as with systematic changes in output. Natural regeneration of vegetation may or may not occur after the land is abandoned, depending on the extent of soil erosion that has taken place.

8. Until 1986 the exchange rate was held nearly constant, in the range of 0.8 to 1.0 naira = $1.0, despite a high domestic inflation rate and falling oil prices. It is beyond the scope of this book to go into the intricacies of

estimating the shadow exchange rate (the reciprocal of the standard conversion factor) that prevailed in May 1985, when the above financial costs were estimated; suffice it to say that it was widely agreed that the s.c.f. was then about 0.35. The slightly lower value of the local conversion factor for the arid zone, which was estimated directly from border price data, represents the differential effects of tariffs and other factors on local agriculture. Again, see Ray (1984) for the principles of estimating border prices, conversion factors, and so on. As observed in note 1, the 1986 devaluation of the naira does not significantly affect the analysis.

9. As noted earlier, it is believed that only about 10 percent of the farm families can be reached by the extension services in the first phase of an afforestation program for the region.

10. About 3–5 percent of the area may have to be taken out of production if farm forestry is to have a large effect on wind velocity. Fairly small effects are assumed in this study, however, and until the trees reach maturity the area taken out of production is not large (less than 2 percent). The effects on the results of using a 3–5 percent figure can be readily estimated.

CHAPTER 5

# Cost-Benefit Analysis: Results and Implications

THE NET BENEFITS OF AFFORESTATION, both for shelterbelts and for farm forestry, are the present value of

- The changes in net farm income from cropping and livestock activities (corresponding to benefits 1, 2, and 3 in chapter 4)
- Plus the benefits of wood and fruit production
- Minus program costs.

Below, this relation is examined under the range of assumptions discussed in chapter 4, with the additional assumption that the costs of afforestation might be higher than estimated. The formulas used in the analysis are presented in the appendix to the book.

## Results

The above relation can be broken down into six terms.[1] (See figures 2 and 3 and the calculations in tables 11 and 12, below.)

1. *Gross farm income with the project* starts out at 200 naira per hectare on the area not occupied by the trees. Soil fertility is taken to decline steadily (at 0 to 2 percent a year, depending on the case) until the trees begin to mature and the decline is slowly stemmed. At this point yields increase and so does the gross value of farm output, for the reasons described in chapter 4, including an increased supply of stover and browse (and thus an increase in gross income from livestock). The benefits of wood production and, in the case of farm forestry, of fruit production also appear then and are included in gross farm income. Beginning in year 15 there is a further upward trend in gross and net farm incomes owing to technical progress in agriculture (see table 9, row 9).

An issue that has not been resolved empirically is whether the increase in gross income associated with the ecological effects of the project is relative to the further degraded soil conditions at the time the afforestation programs are beginning to have their effect (soil fertility is assumed to decline steadily from its initial level) or to the initial soil conditions. In the first case the incremental yield effect amounts to slightly more than restoration of soil fertility to its initial condition; in the second case it amounts to restoration of the soil to the initial condition plus the incremental yield effect and obviously leads to higher benefits. Unfortunately, the available data on the effects of afforestation are not suitable for deciding which case is appropriate because conditions in the control areas are not sufficiently well documented in the studies on shelterbelts cited in chapter 4. In the following analysis the yield effect is first assumed to be relative to further degraded soil conditions; the sensitivity studies then consider the restorative and incremental effects together, since this is a possible outcome.

2. *Net farm income with the project* is gross farm income (term 1) minus farm costs. Farm costs are taken to be 85 percent of the initial gross farm income and are held constant until agricultural productivity begins to rise autonomously in year 15 under the influences described above. The yield benefits arising from the ecological effects of the trees have a negligible effect on costs,

Figure 2. *Shelterbelts (Base Case): Gross and Net Farm Incomes with and without the Project*

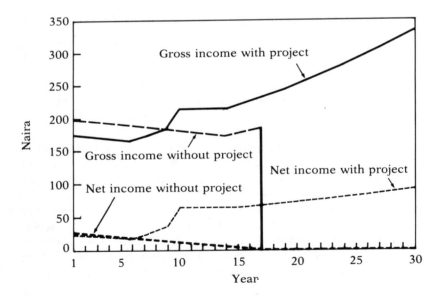

since they are akin to windfall benefits—in this case the benefits arise from improvements in soil moisture and nutrient content. Before the tree planting programs take effect, soil fertility and thus gross farm income might decline while farm costs do not. The possibility then arises that net farm income will become negative and the land will be abandoned; this is allowed for in the model.

3. *Gross farm income without the project* is initially higher than in the case with the project because the land occupied by the shelterbelts would otherwise have been farmed. Income declines directly with the decline in soil fertility but could increase later with the autonomous increase in agricultural productivity if the land has not been abandoned by then.[2]

4. *Net farm income without the project* equals gross farm income (term 3) minus farm costs. As in the with-project case, costs do not decline with soil fertility, and land is taken out of production when net farm income is zero or negative.

5. *Project impact* equals net farm income with the project, minus net farm income without it, minus the project costs. The costs of maintaining the trees and harvesting their production of

**Figure 3.** *Farm Forestry (Base Case): Gross and Net Farm Incomes with and without the Project*

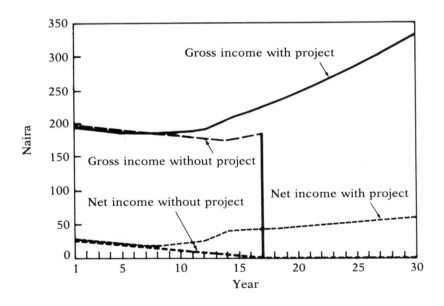

wood, fruit, and fodder are already netted out of the value of their products; hence project costs include only the costs to the project and to the farmers of planting and protecting the trees.

6. *Discounted value of project impact* equals the PV of term 5.

The estimated net benefits and rates of return to investment are shown in table 10. Estimates of the benefits of wood and fruit production were also made separately by putting the ecological benefits at zero to see what difference the latter make in the overall returns. The remarks in the last column of table 10 explain the assumptions used.

Figures 2 and 3 illustrate the base cases for shelterbelts and farm forestry, respectively, and tables 11 and 12 present the calculations in more detail. Consider first shelterbelts. The two upper curves in figure 2 show gross farm income with and without the project, the two lower curves, net farm income. At first both gross and net farm incomes with the shelterbelts are lower than without the project because the shelterbelts take about 12 percent of the land out of production. In both situations, however, gross and net incomes decline over time with the gradual declines in soil fertility, but with the shelterbelts the decline is gradually stopped and is then followed by increases in farm output and wood production as the trees reach maturity. Without the project the declines in soil fertility and thus in gross and net incomes continue. When net income declines to zero without the project, the land is abandoned and gross income also becomes zero. With the project, in contrast, a long-term rise in gross and net incomes eventually occurs because of technical progress in agriculture.

The curves for farm forestry in figure 3 are similar, the main difference being that the buildup of benefits is slower and the yield effects are smaller. But the proportion of land taken out of production by the trees is also smaller.

The technical assumptions on which the above calculations were based are conservative. Yet the prospective economic returns—assuming, in the case of farm forestry, that a good response by farmers can be achieved—are extraordinarily high, around 15 percent for shelterbelts, plus or minus 2 percent, and 19 percent for farm forestry, plus or minus 3 percent. The prospective returns for wood and fruit are modest, about 4–7 percent, but the favorable effects on cropping and livestock activities bring the prospective returns to the orders estimated and possibly higher. The costs of the programs are small in relation to the gross and net values of the cropping and livestock activities protected—

around $10 per hectare a year for shelterbelts and $2 per hectare a year for farm forestry (taking a 10 percent discount rate and netting out the benefits of wood and fruit), or only one-twentieth and one-hundredth, respectively, of the gross value of farm output. Since by arresting declines in soil fertility and enhancing farm output the ecological effects raise gross production without increasing costs—and might, in fact, reduce costs by reducing crop damage during storms—small effects greatly influence the economic return.

Such calculations point to the crucial importance of obtaining a good response by farmers to farm forestry programs. Shelterbelts have prospects of good and more certain returns but significantly higher costs. Farm forestry has prospects of significantly better returns but greater risks. Throughout the analysis the technical assumptions consistently err on the conservative side. The main risks, however, lie not in the technological or economic assumptions but in the social, economic, institutional, and tenurial factors that affect how farmers will respond to the programs. The risks were examined in chapter 3 in light of the experience gained under the 1979–84 afforestation program (AZAP). It was concluded that to improve participation by farmers it is necessary to involve the agricultural extension service and regularly monitor farmer response (neither of which was done under AZAP), and to concentrate on well-defined areas instead of trying for immediate regionwide coverage.

A seeming anomaly in the sensitivity analysis presented in table 10 is that the benefits are lower when the rates of erosion rise from 1 to 2 percent a year (see shelterbelts, case 4, and farm forestry, case 3). The reason is that at higher rates of erosion the value of enhancing the fertility of an increasingly degraded soil is much less. For example, a 20 percent enhancement of yields when soil fertility has already been degraded by 20 percent is much less valuable (in relation to the without-project case) than when soil fertility has been degraded by, say, 10 percent. This raises an implication for policy and a problem for research. The policy implication is that it is generally advisable to protect the better soils before the threat of rapid degradation is serious; that is, to do the projects sooner rather than later. The question for research is not only whether tree planting programs arrest the decline in soil fertility and then enhance fertility, from a degraded condition, by the orders shown (15–25 percent for shelterbelts) but also whether they restore soil fertility to its initial condition and then achieve

Table 10. *Results of Cost-Benefit Studies*

| Case[a] | Yield effect (percent) | Costs relative to base case (percent) | Rate of decline of soil fertility (percent) | Cost-benefit results | | | Remarks |
|---|---|---|---|---|---|---|---|
| | | | | NPV (naira per hectare farmed)[b] | B÷C | IRR (percent) | |
| **Shelterbelts** | | | | | | | |
| 1 | 20 | 100 | 1 | 170 | 2.2 | 14.9 | Base case |
| 2 | 15 | 110 | 1 | 110 | 1.7 | 13.1 | Low yield/high cost case |
| 3 | 25 | 100 | 1 | 221 | 2.6 | 16.2 | High yield case |
| 4 | 20 | 100 | 0 | 108 | 1.8 | 13.5 | No erosion |
| 5 | 20 | 100 | 2 | 109 | 1.8 | 13.6 | More rapid erosion |
| 6 | 20+ | 100 | 1 | 263 | 2.9 | 16.9 | Soil restored to initial condition, plus yield jump |
| 7 | 0 | 100 | 0 | −95 | 0.3 | 4.7 | Wood benefits only |
| **Farm forestry** | | | | | | | |
| 1 | 10 | 100 | 1 | 129 | 4.5 | 19.1 | Base case |
| 2 | 5 | 150[c] | 1 | 70 | 2.3 | 14.5 | Low case (no "high" case assumed) |
| 3 | 10 | 100 | 0 | 75 | 2.9 | 16.6 | No erosion |
| 4 | 10 | 100 | 2 | 60 | 2.5 | 15.5 | More rapid erosion |
| 5 | 10+ | 100 | 1 | 203 | 6.1 | 21.8 | Soil restored to initial condition, plus yield jump |
| 6 | 0 | 100 | 0 | −14 | 0.6 | 7.4 | Wood and fruit benefits only |

*Note:* B, benefits; C, costs; NPV, net present value; IRR, internal rate of return.

a. See tables 11 and 12 for full printouts of the results of the base cases.

b. A 10 percent discount rate was used.

c. This increase corresponds to a three-to-four-year lag in farmer response, plus a 10 percent cost increase.

the yield jumps (shelterbelts, case 6, and farm forestry, case 5). Some foresters and agronomists think that the latter may apply, in which case, the rewards of afforestation could be appreciable indeed.

## Economic Analysis and Scientific Research on Afforestation

The preceding analysis has revealed widespread deficiencies in empirical data on soil conditions in the region, on changes in soil conditions over time, and on the effectiveness of farm trees and shelterbelts in raising the moisture and nutrient content of the soil, reducing erosion, lessening crop damage from storms, and, in general, raising farm output. The assumptions used in the analysis have been reviewed by foresters, agronomists, and others and are generally considered conservative. There is no good substitute, however, for gathering evidence so that a better assessment of the prospective effects of shelterbelts and farm trees on crop yields can be made. Five topics of importance are:

1. Crop yields in the sheltered areas of already established shelterbelts, compared with crop yields in control areas. Such measurements ought to be undertaken for several years; of special interest is the performance of shelterbelts in dry years.

2. Analysis of soil conditions, soil erosion, and the influence of agroforestry and shelterbelts on these quantities. Of particular interest, given the dearth of information on the subject, is the influence of the more dispersed plantings of farm trees on surface wind velocities and on retention of soil moisture. Although controlled experiments (and wind tunnel experiments) on this subject could be attempted, the possibilities of applying multivariate analysis to data gathered from areas with differing tree densities would also be worth exploring. For example, suppose we wish to explain a variable such as soil moisture content, local average wind velocity, or crop yield. Denote the variable by $Z_j$, where $j$ is the area where the variable is measured ($j = 1 \ldots J$). Then it may be possible to estimate functions such as $Z_j = f(D_j; X_j \ldots X_{mj})$, where $d_j$ is the local tree density and $X_j \ldots X_{mj}$ denote all other postulated explanatory variables. The approach would obviously depend on the identification of areas with varying tree densities, but it could complement controlled experiments in a useful way.

Table 11. *Shelterbelts (Base Case): Changes in Farm Incomes with and without Afforestation Investments* (naira, unless otherwise indicated)

| Year | Ecological benefits ($S_1$) | Productivity growth ($S_2$) | Wood benefits ($S_3$) | Farm income per hectare with project Gross | Net | Farm income per hectare without project Gross | Net | Project impact | Discounted value |
|---|---|---|---|---|---|---|---|---|---|
| 1 | 0 | 0 | 0 | 174.2 | 24.6 | 198.0 | 28.0 | -153.4 | -139.4 |
| 2 | 0 | 0 | 0 | 172.5 | 22.9 | 196.0 | 26.0 | -3.1 | -2.6 |
| 3 | 0 | 0 | 0 | 170.8 | 21.2 | 194.1 | 24.1 | -2.9 | -2.2 |
| 4 | 0 | 0 | 0 | 169.1 | 19.5 | 192.2 | 22.2 | -2.7 | -1.8 |
| 5 | 0 | 0 | 0 | 167.4 | 17.8 | 190.2 | 20.2 | -2.4 | -1.5 |
| 6 | 0 | 0 | 0 | 165.8 | 16.2 | 188.4 | 18.4 | -2.2 | -1.2 |
| 7 | 0.2 | 0 | 0 | 170.3 | 20.7 | 186.5 | 16.5 | 4.2 | 2.2 |
| 8 | 0.5 | 0 | 0 | 178.7 | 29.1 | 184.6 | 14.6 | 14.5 | 6.8 |
| 9 | 0.7 | 0 | 0 | 184.7 | 35.1 | 182.8 | 12.8 | 22.3 | 9.4 |
| 10 | 1 | 0 | 1 | 214.3 | 64.7 | 181.0 | 11.0 | 53.8 | 20.7 |
| 11 | 1 | 0 | 1 | 214.3 | 64.7 | 179.2 | 9.2 | 55.6 | 19.5 |
| 12 | 1 | 0 | 1 | 214.3 | 64.7 | 177.4 | 7.4 | 57.3 | 18.3 |
| 13 | 1 | 0 | 1 | 214.3 | 64.7 | 175.6 | 5.6 | 59.1 | 17.1 |
| 14 | 1 | 0 | 1 | 214.3 | 64.7 | 173.9 | 3.9 | 60.9 | 16.0 |
| 15 | 1 | 1 | 1 | 220.3 | 66.1 | 177.4 | 2.2 | 63.9 | 15.3 |
| 16 | 1 | 1 | 1 | 226.4 | 67.5 | 181.0 | 0.5 | 67.1 | 14.6 |
| 17 | 1 | 1 | 1 | 232.7 | 69.0 | 0.0 | 0.0 | 69.0 | 13.7 |
| 18 | 1 | 1 | 1 | 239.2 | 70.5 | 0.0 | 0.0 | 70.5 | 12.7 |
| 19 | 1 | 1 | 1 | 245.9 | 72.1 | 0.0 | 0.0 | 72.1 | 11.8 |
| 20 | 1 | 1 | 1 | 252.8 | 73.7 | 0.0 | 0.0 | 73.7 | 11.0 |
| 21 | 1 | 1 | 1 | 259.9 | 75.3 | 0.0 | 0.0 | 75.3 | 10.2 |
| 22 | 1 | 1 | 1 | 267.2 | 77.0 | 0.0 | 0.0 | 77.0 | 9.5 |
| 23 | 1 | 1 | 1 | 274.8 | 78.8 | 0.0 | 0.0 | 78.8 | 8.8 |

| | | | | | | | | |
|---|---|---|---|---|---|---|---|---|
| 24 | 1 | 1 | 282.5 | 0.0 | 0.0 | 80.6 | 80.6 | 8.2 |
| 25 | 1 | 1 | 290.5 | 0.0 | 0.0 | 82.5 | 82.5 | 7.6 |
| 26 | 1 | 1 | 298.8 | 0.0 | 0.0 | 84.4 | 84.4 | 7.1 |
| 27 | 1 | 1 | 307.3 | 0.0 | 0.0 | 86.4 | 86.4 | 6.6 |
| 28 | 1 | 1 | 316.1 | 0.0 | 0.0 | 88.4 | 88.4 | 6.1 |
| 29 | 1 | 1 | 325.1 | 0.0 | 0.0 | 90.5 | 90.5 | 5.7 |
| 30 | 1 | 1 | 334.4 | 0.0 | 0.0 | 92.7 | 92.7 | 5.3 |
| 31 | 1 | 1 | 344.0 | 0.0 | 0.0 | 94.9 | 94.9 | 4.9 |
| 32 | 1 | 1 | 353.9 | 0.0 | 0.0 | 97.2 | 97.2 | 4.6 |
| 33 | 1 | 1 | 364.1 | 0.0 | 0.0 | 99.6 | 99.6 | 4.3 |
| 34 | 1 | 1 | 374.6 | 0.0 | 0.0 | 102.0 | 102.0 | 4.0 |
| 35 | 1 | 1 | 385.4 | 0.0 | 0.0 | 104.5 | 104.5 | 3.7 |
| 36 | 1 | 1 | 396.6 | 0.0 | 0.0 | 107.1 | 107.1 | 3.5 |
| 37 | 1 | 1 | 408.1 | 0.0 | 0.0 | 109.8 | 109.8 | 3.2 |
| 38 | 1 | 1 | 419.9 | 0.0 | 0.0 | 112.6 | 112.6 | 3.0 |
| 39 | 1 | 1 | 432.1 | 0.0 | 0.0 | 115.4 | 115.4 | 2.8 |
| 40 | 1 | 1 | 444.7 | 0.0 | 0.0 | 118.3 | 118.3 | 2.6 |
| 41 | 1 | 1 | 457.6 | 0.0 | 0.0 | 121.3 | 121.3 | 2.4 |
| 42 | 1 | 1 | 471.0 | 0.0 | 0.0 | 124.4 | 124.4 | 2.3 |
| 43 | 1 | 1 | 484.7 | 0.0 | 0.0 | 127.6 | 127.6 | 2.1 |
| 44 | 1 | 1 | 498.9 | 0.0 | 0.0 | 130.9 | 130.9 | 2.0 |
| 45 | 1 | 1 | 513.5 | 0.0 | 0.0 | 134.3 | 134.3 | 1.8 |
| 46 | 1 | 1 | 528.5 | 0.0 | 0.0 | 137.8 | 137.8 | 1.7 |
| 47 | 1 | 1 | 544.0 | 0.0 | 0.0 | 141.4 | 141.4 | 1.6 |
| 48 | 1 | 1 | 560.0 | 0.0 | 0.0 | 145.2 | 145.2 | 1.5 |
| 49 | 1 | 1 | 576.5 | 0.0 | 0.0 | 149.0 | 149.0 | 1.4 |
| 50 | 1 | 1 | 593.5 | 0.0 | 0.0 | 152.9 | 152.9 | 1.3 |
| NPV | — | — | — | — | — | — | — | 170.1 |

— Not applicable or not calculated.   *Note:* See tables 9 and 10 for assumptions. NPV, net present value.

Table 12. *Farm-Forestry (Base Case): Changes in Farm Incomes with and without Afforestation Investments* (naira, unless otherwise indicated)

| Year | Dummy variables | | | Farm income per hectare with project | | Farm income per hectare without project | | Project impact | Discounted value |
|---|---|---|---|---|---|---|---|---|---|
| | Ecological benefits ($S_1$) | Productivity growth ($S_2$) | Wood benefits ($S_3$) | Gross | Net | Gross | Net | | |
| 1 | 0 | 0 | 0 | 194.0 | 27.4 | 198.0 | 28.0 | -40.6 | -36.9 |
| 2 | 0 | 0 | 0 | 192.1 | 25.5 | 196.0 | 26.0 | -0.5 | -0.4 |
| 3 | 0 | 0 | 0 | 190.2 | 23.6 | 194.1 | 24.1 | -0.5 | -0.4 |
| 4 | 0 | 0 | 0 | 188.3 | 21.7 | 192.2 | 22.2 | -0.4 | -0.3 |
| 5 | 0 | 0 | 0 | 186.4 | 19.8 | 190.2 | 20.2 | -0.4 | -0.3 |
| 6 | 0 | 0 | 0.1 | 185.3 | 18.7 | 188.4 | 18.4 | 0.3 | 0.2 |
| 7 | 0 | 0 | 0.2 | 184.1 | 17.5 | 186.5 | 16.5 | 1.0 | 0.5 |
| 8 | 0.1 | 0 | 0.3 | 184.8 | 18.2 | 184.6 | 14.6 | 3.6 | 1.7 |
| 9 | 0.2 | 0 | 0.4 | 185.8 | 19.2 | 182.8 | 12.8 | 6.4 | 2.7 |
| 10 | 0.3 | 0 | 0.5 | 187.2 | 20.6 | 181.0 | 11.0 | 9.6 | 3.7 |
| 11 | 0.4 | 0 | 0.6 | 188.9 | 22.3 | 179.2 | 9.2 | 13.2 | 4.6 |
| 12 | 0.5 | 0 | 0.8 | 191.7 | 25.1 | 177.4 | 7.4 | 17.7 | 5.6 |
| 13 | 0.7 | 0 | 1 | 197.6 | 31.0 | 175.6 | 5.6 | 25.4 | 7.3 |
| 14 | 1 | 0 | 1 | 205.9 | 39.3 | 173.9 | 3.9 | 35.4 | 9.3 |
| 15 | 1 | 1 | 1 | 211.9 | 40.3 | 177.4 | 2.2 | 38.1 | 9.1 |
| 16 | 1 | 1 | 1 | 218.2 | 41.3 | 181.0 | 0.5 | 40.8 | 8.9 |
| 17 | 1 | 1 | 1 | 224.6 | 42.3 | 0.0 | 0.0 | 42.3 | 8.4 |
| 18 | 1 | 1 | 1 | 231.3 | 43.4 | 0.0 | 0.0 | 43.4 | 7.8 |
| 19 | 1 | 1 | 1 | 238.1 | 44.5 | 0.0 | 0.0 | 44.5 | 7.3 |
| 20 | 1 | 1 | 1 | 245.1 | 45.7 | 0.0 | 0.0 | 45.7 | 6.8 |
| 21 | 1 | 1 | 1 | 252.4 | 46.9 | 0.0 | 0.0 | 46.9 | 6.3 |
| 22 | 1 | 1 | 1 | 259.9 | 48.1 | 0.0 | 0.0 | 48.1 | 5.9 |
| 23 | 1 | 1 | 1 | 267.6 | 49.3 | 0.0 | 0.0 | 49.3 | 5.5 |

| | | | | | | | | | | |
|---|---|---|---|---|---|---|---|---|---|---|
| 24 | 1 | 1 | 1 | 1 | 275.5 | 50.6 | 0.0 | 0.0 | 50.6 | 5.1 |
| 25 | 1 | 1 | 1 | 1 | 283.7 | 52.0 | 0.0 | 0.0 | 52.0 | 4.8 |
| 26 | 1 | 1 | 1 | 1 | 292.1 | 53.3 | 0.0 | 0.0 | 53.3 | 4.5 |
| 27 | 1 | 1 | 1 | 1 | 300.8 | 54.7 | 0.0 | 0.0 | 54.7 | 4.2 |
| 28 | 1 | 1 | 1 | 1 | 309.8 | 56.2 | 0.0 | 0.0 | 56.2 | 3.9 |
| 29 | 1 | 1 | 1 | 1 | 319.0 | 57.7 | 0.0 | 0.0 | 57.7 | 3.6 |
| 30 | 1 | 1 | 1 | 1 | 328.5 | 59.3 | 0.0 | 0.0 | 59.3 | 3.4 |
| 31 | 1 | 1 | 1 | 1 | 338.3 | 60.9 | 0.0 | 0.0 | 60.9 | 3.2 |
| 32 | 1 | 1 | 1 | 1 | 348.4 | 62.5 | 0.0 | 0.0 | 62.5 | 3.0 |
| 33 | 1 | 1 | 1 | 1 | 358.8 | 64.2 | 0.0 | 0.0 | 64.2 | 2.8 |
| 34 | 1 | 1 | 1 | 1 | 369.5 | 65.9 | 0.0 | 0.0 | 65.9 | 2.6 |
| 35 | 1 | 1 | 1 | 1 | 380.5 | 67.7 | 0.0 | 0.0 | 67.7 | 2.4 |
| 36 | 1 | 1 | 1 | 1 | 391.9 | 69.6 | 0.0 | 0.0 | 69.6 | 2.3 |
| 37 | 1 | 1 | 1 | 1 | 403.7 | 71.5 | 0.0 | 0.0 | 71.5 | 2.1 |
| 38 | 1 | 1 | 1 | 1 | 415.7 | 73.5 | 0.0 | 0.0 | 73.5 | 2.0 |
| 39 | 1 | 1 | 1 | 1 | 428.2 | 75.5 | 0.0 | 0.0 | 75.5 | 1.8 |
| 40 | 1 | 1 | 1 | 1 | 441.0 | 77.6 | 0.0 | 0.0 | 77.6 | 1.7 |
| 41 | 1 | 1 | 1 | 1 | 454.2 | 79.7 | 0.0 | 0.0 | 79.7 | 1.6 |
| 42 | 1 | 1 | 1 | 1 | 467.9 | 82.0 | 0.0 | 0.0 | 82.0 | 1.5 |
| 43 | 1 | 1 | 1 | 1 | 481.9 | 84.3 | 0.0 | 0.0 | 84.3 | 1.4 |
| 44 | 1 | 1 | 1 | 1 | 496.4 | 86.6 | 0.0 | 0.0 | 86.6 | 1.3 |
| 45 | 1 | 1 | 1 | 1 | 511.3 | 89.0 | 0.0 | 0.0 | 89.0 | 1.2 |
| 46 | 1 | 1 | 1 | 1 | 526.6 | 91.5 | 0.0 | 0.0 | 91.5 | 1.1 |
| 47 | 1 | 1 | 1 | 1 | 542.5 | 94.1 | 0.0 | 0.0 | 94.1 | 1.1 |
| 48 | 1 | 1 | 1 | 1 | 558.8 | 96.8 | 0.0 | 0.0 | 96.8 | 1.0 |
| 49 | 1 | 1 | 1 | 1 | 575.6 | 99.5 | 0.0 | 0.0 | 99.5 | 0.9 |
| 50 | 1 | 1 | 1 | 1 | 592.9 | 102.3 | 0.0 | 0.0 | 102.3 | 0.9 |
| NPV | — | — | — | — | — | — | — | — | — | 128.8 |

— Not applicable or not calculated.    *Note:* See tables 9 and 10 for assumptions. NPV, net present value.

Indeed, the procedures of multivariate analysis could be used to pool these different data sources.

3. A study of the influence of farming systems—in particular of agropastoral practices—on the returns to shelterbelts and farm forestry; conversely, given the greater availability of green mulch, fodder, and dung, a study of the influence of shelterbelts and farm forestry on agropastoral practices.

4. A study of the optimal spacing of shelterbelts and (given the complementarities between the two) of the most productive ways of combining the shelterbelt and farm forestry programs in afforestation plans.

5. Similarly, a study of optimal farm tree densities and species. Studies which consider this question in general terms (for example, Raintree 1984) were cited in chapter 2, but there is no good substitute for local measurement and analysis tailored to the local system.

In addition (this is a general point for the arid regions of Africa), there are good arguments for the development of improved technological packages, for breeding programs using survivor trees in drought areas, and, more generally, for scientific research on agro-silvo-pastoral associations (see chapter 2).

All of the above suggestions relate to the technical design of afforestation programs. But, as discussed in chapters 1 and 2, there is much to be learned about the social and economic causes of the changes now taking place. For example, hypotheses about desertification abound: among the partial explanations cited are the "commons problem," the problem of property rights, the problem of downwind and downstream (external) effects, the bias against agriculture and rural areas in economic policy, the propensity of low-income groups to discount the future highly, mechanization, and even ignorance. There is an obvious need for research by social scientists to clarify these matters and (equally important) to draw out the policy implications of the findings. Another socioeconomic question worth addressing is whether the price mechanism can be made to work better—for example, by taxing farmers for not having trees or by subsidizing them for having them. Such an approach, although fraught with administrative difficulties and the danger that funds might be misappropriated, is nevertheless worth analysis.

The attempt to determine the sources of ecological deterioration and desertification in farming areas, however, cannot be limited to social, economic, or political considerations alone.

Such factors must be related to the measurement and analysis of the ecological changes taking place. Consider, for instance, the highly suggestive analysis by Newcombe (1984) of the desertification process in Ethiopia, which well indicates the range of economic and ecological variables involved. He notes that the depletion of the tree cover and the subsequent use of animal dung and crop residues for fuel breach all three of the natural cycles through which nutrients are returned to an initially rich topsoil: (1) the tree cycle, (2) the grass and crop residue cycle, and (3) the dung conversion cycle. The process of desertification is divided into five phases:

• *Phase I.* Deterioration in the ecology begins when the rate of timber harvesting for all purposes exceeds the average annual rate of production. This may occur because of major land clearing projects or with the growth of demand for fuelwood and charcoal by local urban and rural households. Nutrient cycle 1, the tree cycle, begins to decline, with an at first imperceptible impact on food production.

• *Phase II.* As the demand for fuelwood grows and wood becomes scarce, peasants begin to use cereal straw and dung for fuel. Nutrient cycles 2 and 3, the crop residue and dung cycles, are breached for the first time and the recycling of nutrients diminishes. Use of crop residues and dung for cooking leads to lower inputs of organic matter, poor soil structure, lower retention of nutrients in the crop root zone, and reduced protection from the erosive effects of wind and heavy rains. The nutrient reserves of the topsoil begin to decline.

• *Phase III.* Virtually all tree cover is gone and the tree cycle ceases completely. Now a high proportion of the dung produced is collected, woody cereal stalks are systematically stored, and both are sold for cash to urban markets. The cereal crop yields and, in consequence, the animal carrying capacity begin to decline, and the land becomes overgrazed. Soil erosion becomes serious.

• *Phase IV.* Dung is the only source of fuel and has become a major source of income. All crop residues are used for animal feed, but they are not sufficient for the purpose. Nutrient cycle 2 is negligible and nutrient cycle 3 is greatly reduced. Soil erosion is dramatic and the topsoil is much depleted. In such a situation extended dry periods can be devastating because the ecosystem loses its capacity to recover quickly.

• *Phase V.* Production of organic matter undergoes a total collapse, usually catalyzed by dry periods, which were previously tolerable. Peasants abandon their land, which has now become desert.

Newcombe adds that "this transition from the first to the final stage is in process right across Ethiopia and has reached the terminal phase in parts of Tigrai and Eritrea."

It should be emphasized that there is nothing inevitable about the above process, which has been brought about as much by political and economic events as by any other factor. It is also a without-project scenario. On a more encouraging note Timberlake (1985, pp. 136–40) describes an apparently successful terracing and afforestation project undertaken by farmers in Ethiopia's Central Highlands which has done much to halt erosion and stabilize crop yields in the area. The experiences of India, China, and other countries also offer grounds for encouragement (see notes 5 and 6 in chapter 2).

## Notes

1. This way of formulating the calculations was proposed by my colleague Kenneth Sigrist. It is formulated mathematically in the appendix.

2. The model presented in the appendix permits the land to be brought back into production in the without-project case if technical progress is sufficient to offset the losses in soil fertility and give positive net incomes. This, however, did not occur under any of the assumptions examined above.

# Conclusion

AS THE CASE STUDY IN PART II has sought to illustrate, workable approaches for addressing the ecological problems caused by land clearing and fuelwood consumption in Africa do exist. Despite the complexity and diversity of ecosystems and of regional economic conditions, there are common elements in the policies put forward by agronomists and foresters. Of these approaches, farm forestry and shelterbelt and watershed plantings hold much promise. The case study in chapters 3–5 sought to show that the prospective returns to such investments may be large once the ecological benefits are taken into account and if—this remains a strong qualification—farmers respond well to the programs. Participation by the farmers depends on favorable economic and tenurial incentives for investment in and maintenance of land resources. Such incentives are, however, only necessary, not sufficient, conditions, given the weakness of forestry laws in the region and the short-term private gains from tree cutting. Because destruction of trees ultimately affects the common welfare, public investment in afforestation should be accorded a high economic priority in the ecologically threatened areas of Africa.

In addition to direct investment in and public support for afforestation, the preceding analysis has underscored the need for research on several subjects—on changes in soil fertility, on technology packages, on the responses of soils (and thus of farm output) to afforestation programs, and on the social, tenurial, and economic factors that affect the farmers' response to afforestation programs. Suggestions for research were put forward in chapter 5.

There is widespread agreement among agronomists and foresters on the ecological and economic importance of investments in afforestation and of the associated improvements in land management practices. But the recommended policies have not been applied on a significant scale in Africa because public recognition of the problems and a commitment to addressing them have been lacking. Part of the emerging tragedy is that the resources required would be small in relation to the prospective economic gains from wood production and from rising (as opposed to declining) soil fertility.

APPENDIX

# Cost-Benefit Analysis: Mathematical Background

THIS APPENDIX SETS OUT the formulas used for the cost-benefit calculations presented in chapter 5. The formulas are for gross farm income and net farm income with the project, gross farm income and net farm income without the project, project impact (the change in net farm income less project costs), and the discounted value of project impact.

1. *Gross farm income with the project* $(B_t)$. Let $Y$ denote gross farm income in year 0 and $y$ the proportional increase in gross farm income arising from the ecological benefits of the project. In the absence of other changes affecting farm output (such as declining soil fertility and changes in agricultural productivity), the gross farm income per hectare in any year $t$ is given by

(1) $$B_t = a_1 Y(1 + yS_1) + a_1 WS_3$$

where $a_1$ is the fraction of the hectare that is farmed, $(1 - a_1$ is thus the fraction taken out of production by the trees), $S_1$ is a dummy variable representing the buildup of the ecological benefits, $W$ denotes the wood benefits, and $S_3$ is a dummy variable representing the buildup of the wood benefits. (The variables are also listed in table 13.) The values of $S_1$ and $S_3$ (from the tables in chapter 4) are as follows.

| | 1-5 | 6 | 7 | 8 | 9 | 10 | 11 | 12 | 13 | 14 | 15+ |
|---|---|---|---|---|---|---|---|---|---|---|---|
| Shelterbelts | | | | | | | | | | | |
| $S_1$ | 0.0 | 0.0 | 0.2 | 0.5 | 0.7 | 1.0 | 1.0 | 1.0 | 1.0 | 1.0 | 1.0 |
| $S_3$ | 0.0 | 0.0 | 0.0 | 0.0 | 0.0 | 1.0 | 1.0 | 1.0 | 1.0 | 1.0 | 1.0 |
| Farm forestry | | | | | | | | | | | |
| $S_1$ | 0.0 | 0.0 | 0.0 | 0.1 | 0.2 | 0.3 | 0.4 | 0.5 | 0.7 | 1.0 | 1.0 |
| $S_3$ | 0.0 | 0.1 | 0.2 | 0.3 | 0.4 | 0.5 | 0.6 | 0.8 | 1.0 | 1.0 | 1.0 |

The first term on the right-hand side of equation 1 needs to be modified to allow for two factors: a gradual decline in soil fertility at a rate of $d$ a year until the tree plantings begin to stabilize soil conditions (in, say, year $T_1$) and the autonomous rise in agricul-

79

Table 13.   *Variables Used in the Cost-Benefit Analysis*

| Variable | Definition |
| --- | --- |
| $a$ | Ratio of farm costs to farm income in year 0 |
| $a_1$ | Fraction of area farmed |
| $1 - a_1$ | Fraction of area occupied by trees |
| $B_t$ | Gross farm income with project in year $t$ |
| $b_t$ | Gross farm income without project in year $t$ |
| $C_t$ | Project costs in year $t$ |
| $D_t$ | Discounted value of project impact in year $t$ |
| $d$ | Rate of decline in soil fertility per year |
| $g$ | Autonomous rise in agricultural productivity per year |
| $I_t$ | Project impact in year $t$ |
| $N_t$ | Net farm income with project in year $t$ |
| $n_t$ | Net farm income without project in year $t$ |
| $S_1$ | Dummy: buildup of ecological benefits |
| $S_2$ | Dummy: growth of agricultural productivity |
| $S_3$ | Dummy: buildup of wood benefits |
| $T$ | Time (year) |
| $W$ | Wood benefits |
| $Y$ | Gross farm income in year 1 |
| $y$ | Proportional increase in gross farm income arising from ecological benefits |

tural productivity, at a rate of $g$ a year, beginning in, say, year $T_2$. ($T_1$ is taken to be year 8 and $T_2$ to be year 15.) Applying trend variables to equation 1 gives

$$(1a) \quad B_t = a_1 Y (1 + S_1 y) \exp[-d(1 - S_1)t - dS_1 T_1]$$
$$\cdot \exp [S_2(t - T_2)g] + a_1 W S_3.$$

The exponential term $\exp[S_2(t - T_2)g]$ represents productivity growth in agriculture arising from technical change, with $S_2$ being a dummy variable that has the value 0 in years 1 to 14 and 1.0 in year 15 and afterward. The term $[-d(1 - S_1)t - d_1 S_1 T_1]$ in the first exponent represents the gradual stemming and then prevention of the declines in soil fertility in accordance with the values of $S_1$ in the above table. When in the early years $S_1 = 0$ (and $S_2$ and $S_3 = 0$ also),

$$B_t = a_1 Y \exp(-dt)$$

that is, gross output declines steadily. But when $S_1 = 1$, $S_2 = 0$, and $S_3 = 0$,

$$(1b) \quad\quad\quad B_t = a_1 Y (1 + y) \exp(-dT_1).$$

It is apparent from this last expression that the assumed effect of the trees is to increase soil fertility by 100y percent from a degraded condition, since the term $\exp(-dT_1)$ is "permanently" present when $t$ exceeds $T_1$ (which is a defined point in time and is constant). Would the plantings achieve more than this, that is, increase the yields by 100y percent above initial levels? This might be possible in both farm forestry and shelterbelt areas, since the initial (that is, present) tree stocks are already depleted (down to 3 or fewer trees per hectare in much of the arid zone). Then it might be more accurate to drop the term $(-dT_1)$ from the exponent in (1a) and (1b) by putting $T_1$ equal to 0. This was done, to assess what the effects might be, in the sensitivity analysis presented in chapter 5. Unfortunately, the relevant field measurements to support this stronger assumption (recovery of soil yields to the initial condition plus a yield effect of 100y percent) are still lacking. As discussed in chapter 5, a series of control measurements in areas with varying farm tree densities is needed to provide clarification.

2. *Net farm income with the project (N$_t$).* Net farm income is gross farm income minus farm costs. Farm costs are likely to rise with general increases in agricultural productivity, represented by the term $\exp[S_2(t - T_2)g]$ in equation 1a. They are unlikely, however, to rise significantly with the increase in the moisture and nutrient content of the soil or with reductions in crop damage, since these are pure windfall benefits; there may be a slight added cost of harvesting a heavier crop, but this effect is probably negligible. If anything, farm costs are likely to fall in the sheltered areas because crop damage and the frequent need to resow the crops after storms are reduced. Hence the net farm income, $N_t$, with the project is given by

$$(2) \qquad N_t = B_t - a \cdot a_1 \cdot Y \cdot \exp[S_2(t - T_2)g]$$

where $a$ is the ratio of farm costs to gross farm income in the initial year of the project ($a = 0.85$ in this analysis; see chapter 4). It is assumed that if $N_t$ were to become negative (for example, owing to a fall in $B_t$ before the trees have their positive effects), the land would be abandoned. The land could be recovered, however, if the yield jump ($y$), technical progress, or both were sufficient to make $N_t$ positive.

3. *Gross farm income without the project (b$_t$).* The differences between the cases with and without the project are that in the latter case there is no yield jump associated with the tree plantings,

soil fertility declines continuously, and the area farmed is slightly greater, being $a_1 + (1 - a_1) = 1$. Wood and fruit benefits are also zero. Putting $S_1$ and $S_3$ equal to 0 in equation 1a and the area farmed equal to 1.0 gives as the gross farm income in the without-project case

(3) $$b_t = Y \cdot \exp(-dt) \cdot \exp[S_2(t - T_2)g].$$

This expression is zero if net farm income in the without-project case, discussed below, is zero.

In expression 3 technical progress in agriculture, represented by $g$, slows the decline in soil fertility and may arrest it, depending on the relative values of $d$ and $g$. Again, it is a matter of some importance to obtain evidence on these quantities; what the equation represents is steady losses of yields owing to erosion and to declines in the moisture and nutrient content of the soil that occur at the same time as farmers and governmental agencies are attempting to raise yields. In the with-project case, in equation 2, the tree planting programs halt the decline, restore soil fertility, and—of much importance for current rural development programs in Nigeria—make it possible for technological and other innovations to have their full effects on output. The economic cost of erosion and desertification is thus not only much greater than the lost output from present farming practices but also includes the increases in output that might have resulted from technological and institutional innovation with the project.

4. *Net farm income without the project* ($n_t$). This is gross farm income without the project, minus costs, or

(4) $$n_t = b_t - aY \cdot \exp[S_2(t - T_2)g]$$

with land being taken out of production when $n_t$ is negative; that is, $n_t$ is set to zero when the right-hand side of equation 4 is negative.

5. *Project impact* ($I_t$). The impact of the project on farm incomes in year $t$ is the difference between the cases with and without the project minus the project costs, $C_t$. That is,

(5) $$I_t = N_t - n_t - C_t.$$

6. *Discounted value of project impact* ($D_t$). The discounted value of $I_t$ is

(6) $$D_t = I_t(1 + r)^{-t}$$

where $r$ is the discount rate.

# References

Allen, Julia C. 1985. "Wood Energy and Preservation of Woodlands in Semi-Arid Developing Countries: The Case of Dodoma Region, Tanzania." *Journal of Development Economics*, vol. 19 (September–October), pp. 59–84.

Allen, Julia C., and Douglas F. Barnes. 1985. "The Causes of Deforestation in Developing Countries." *Annals of the Association of American Geographers*, vol. 75, no. 2, pp. 163–84.

Anderson, Dennis, and Robert Fishwick. 1984. *Fuelwood Consumption and Deforestation in African Countries: A Review*. World Bank Staff Working Paper 704. Washington, D.C.

Blair, Harry W. 1986. "Social Forestry: A Time to Modify Goals?" *Economic and Political Weekly* (India), vol. 21, no. 30 (July 26), pp. 1317–21.

Bognetteau-Verlinden, Els. 1980. *Study on Impact of Windbreaks in Majjia Valley, Niger*. Wageningen, Netherlands: CARE–Niger Forest Service.

Booth, G. A. 1979. *A Review of Forestry Development in the Arid Zone of Nigeria*. Rome: Food and Agriculture Organization.

Brindle, James R., David Hintz, and J. R. Sturrock. Forthcoming. *Windbreaks*. Proceedings of an International Symposium on Windbreaks, June 1986, University of Nebraska, Lincoln. Amsterdam: Elsevier.

Brown, Lester R., and Edward C. Wolf. 1985. *Reversing Africa's Decline*. Worldwatch Paper 65, Washington, D.C.: Worldwatch Institute.

Burley, Jeffrey. 1982. "Obstacles to Tree Planting in Arid and Semi-Arid Lands: Comparative Case Studies from India and Kenya." NRTS-18/UNDP-391. Tokyo: United Nations University.

Chandler, Trevor, and David Spurgeon. 1979. *Conference on International Co-operation in Agro-Forestry*. Nairobi: German Foundation for International Development (DSE) and the International Development Council for Research in Agro-Forestry.

Clément, Jean. 1982. "Estimation des volumes et de la productivité des formations mixtes forestières et graminéennes tropicales," *Bois et Forets des Tropiques*, no. 198. Centre Technique Forestier Tropical.

Dixon, John A., and Maynard M. Hufschmidt. 1986. *Economic Valuation Techniques for the Environment: A Case Study Workbook*. Baltimore, Md.: Johns Hopkins University Press.

*83*

Easter, K. William, John A. Dixon, and Maynard M. Hufschmidt. 1986. *Watershed Resource Management: An Integrated Framework, with Studies from Asia and the Pacific.* Westview Studies in Water Policy and Management 10. Boulder, Colo.: Westview Press.

Famoriyo, Segun. 1979. *Land Tenure and Agricultural Development in Nigeria,* Nigerian Institute of Social and Economic Research. Ibadan: Ibadan University Press.

Foley, Gerald, and Geoffrey Barnard. 1984. *Farm and Community Forestry.* Earthscan Technical Report 3. London: International Institute for Environment and Development.

Foley, Gerald, Patricia Moss, and Lloyd Timberlake. 1984. *Stoves and Trees.* Earthscan Paperback. London: International Institute for Environment and Development.

Food and Agriculture Organization. 1985. "Tree Growing by Rural People." FAO Forestry Paper 64. Rome.

Fortmann, Louise. 1985. "The Tree Tenure Factor in Agro-forestry with Particular Reference to Africa." *Agro-forestry Systems,* vol. 2, pp. 229–51.

French, David. 1986. "Confronting an Unsolvable Problem: Deforestation in Malawi." *World Development,* vol. 14, no. 4 (April), pp. 531–40.

Fuller, B. F. 1981. "Woodfuel: Planning for Solutions." *Zimbabwe Science News,* vol. 15, no. 1 (January).

Gregersen, H. M., and others. Forthcoming. *Economic Analysis of Watershed Management Projects.* Rome: Food and Agriculture Organization.

Griffin, Keith. 1986a. "Communal Land Tenure Systems and Their Role in Rural Development." In Sanjaya Lall and Frances Stewart, eds. *Theory and Reality in Development: Essays in Honor of Paul Streeten.* New York: St. Martin's Press.

———. 1986b. "Rural Development in an Arid Region: Xinjiang." *Third World Quarterly,* vol. 8, no. 3, pp. 978–1001.

Hill, Polly. 1972. *Rural Hausa: A Village and a Setting.* London: Cambridge University Press.

———. 1986. *Development Economics on Trial.* Cambridge: Cambridge University Press.

Hufschmidt, Maynard M., D. E. James, A. D. Meister, B. T. Bower, and J. A. Dixon. 1983. *Environment, Natural Systems, and Development: An Economic Valuation Guide.* Baltimore, Md.: Johns Hopkins University Press.

Hughart, David P. 1979. *Prospects for Traditional and Non-Conventional Energy Sources in Developing Countries.* World Bank Staff Working Paper 346. Washington, D.C.

Keita, M. N. 1982. "Les Disponibilités de bois de feu en région sahelienne de l'Afrique occidentale: Situation et perspectives." Rome: Food and Agriculture Organization.

Lamprey, H. 1978. "The Integrated Project on Arid Lands (IPAL)." *Nature and Resources* (Unesco), vol. 14, no. 4, pp. 2–11.

Lanly, Jean Paul. 1982. "Tropical Forest Resources." FAO Paper 30. Rome: Food and Agriculture Organization.

Le Houerou, Henri N. 1980. *International Symposium on Browse in Africa: Summary of Contributions.* Addis Ababa: International Livestock Centre for Africa.

Livingstone, Ian. 1986. "The Common Property Problem and Pastoralist Economic Behaviour." *Journal of Development Studies*, vol. 23 (October), 5–19.

Manibog, Fernando R. 1984. "Improved Cooking Stoves in Developing Countries: Problems and Opportunities." *Annual Energy Review*, vol. 9, pp. 199–227.

Marshall, Alfred. 1890/1980. *Principles of Economics.* 8th ed. Philadelphia, Pa.: Porcupine Press.

Mnzava, E. M. 1983. "Tree Planting in Tanzania: A Voice from Villagers." Forest Division, Ministry of Lands, Natural Resources, and Tourism, Tanzania. Dar es Salaam.

Myers, Norman. 1984. *The Primary Source: Tropical Forests and Our Future.* London: Norton.

National Research Council. Board on Science and Technology for International Development. 1984a. *Agroforestry in the West African Sahel.* Washington, D.C.: National Academy Press.

————. 1984b. *Environmental Change in the West African Sahel.* Washington, D.C.: National Academy Press.

Nautiyal, J. C. 1979. "Afforestation Program in the Arid Zone of Nigeria." NIR/75/053, Field Document No. 1. Rome: Food and Agriculture Organization.

Newcombe, Kenneth. 1984. "An Economic Justification for Rural Afforestation: The Case of Ethiopia." Energy Department Paper 16. World Bank, Washington, D.C.

Nigeria, Federal Department of Forestry. 1983. *National Committee on Arid Zone Afforestation: Annual Report.* P.M.B. 2082, Kaduna.

Noronha, Raymond. 1985. "A Review of the Literature on Land Tenure Systems in Sub-Saharan Africa." Agriculture and Rural Development Department Discussion Paper ARU 43. World Bank, Washington, D.C.

Powell, J. Mark. 1984. "Cropping Systems in the Sub-humid Zone." Paper for a Symposium on Livestock, International Livestock Centre for Africa and the National Animal Production Research Institute (ICLA/ NAPRI), Ahmadu Bello University, Zaria, Nigeria, October 30–November 2. Processed.

Raintree, J. B. 1984. "A Diagnostic Approach to Agro-forestry Design." In K. F. Wiersum, ed. *Strategies and Designs for Afforestation, Reforestation, and Tree Planting.* Wageningen, Netherlands: Centre for Agricultural Publishing and Documentation (PUDOC).

Ray, Anandarup. 1984. *Cost-Benefit Analysis: Issues and Methodologies.* Baltimore, Md.: Johns Hopkins University Press.

Runge, Carlisle Ford. 1986. "Common Property and Collective Action in Economic Development." *World Development*, vol. 14, no. 5, pp. 623–35.

Seeley, J. A. 1985. *Conservation in Sub-Saharan Africa: An Introductory Bibliography for the Social Sciences.* Cambridge African Monographs 5. Cambridge, England: Cambridge University Press.

Stebbing, E. P. 1935. "The Encroaching Sahara: The Threat to the West African Colonies." *Geographical Journal*, vol. 85, pp. 508–24.

Tanzania, Ministry of Lands, Natural Resources, and Tourism. 1984. *Trees for Village Forestry.* Dar es Salaam.

Timberlake, Lloyd. 1985. *Africa in Crisis.* Earthscan Paperback. London: International Institute for Environment and Development.

von Maydell, H. J. 1979. "The Development of Agro-forestry in the Sahelian Zone of Africa." In Trevor Chandler and David Spurgeon, eds. *Conference on International Co-operation in Agro-Forestry.* Nairobi: German Foundation for International Development (DSE) and the International Development Council for Research in Agro-Forestry.

———. 1983. *Arbres et arbustes du Sahel: Leurs caractéristiques et leurs utilisations.* Eschborn: Deutsche Gesellschaft für Technische Zusammenarbeit.

Wiersum, K. F. (ed.). 1984. *Strategies and Designs for Afforestation, Reforestation, and Tree Planting.* Wageningen, Netherlands: Centre for Agricultural Publishing and Documentation (PUDOC).

World Bank. 1978. *Forestry Sector Policy Paper.* Washington, D.C.

———. 1981. *Accelerated Development in Sub-Saharan Africa: An Agenda for Action.* Washington, D.C.

———. 1983a. *Commodity Trade and Price Trends.* 1983–84 ed. Washington, D.C.

———. 1983b. *The Energy Transition in Developing Countries.* Washington, D.C.

———. 1987. *World Development Report 1987.* New York: Oxford University Press.

World Resources Institute, World Bank, and United Nations Development Programme. 1985. *Tropical Forests: A Call for Action.* International Task Force Report. Washington, D.C.: World Resources Institute.